New Directions in the Philosophy of Science

Series Editor
Steven French
Department of Philosophy
University of Leeds
Leeds, United Kingdom

The philosophy of science is going through exciting times. New and productive relationships are being sought with the history of science. Illuminating and innovative comparisons are being developed between the philosophy of science and the philosophy of art. The role of mathematics in science is being opened up to renewed scrutiny in the light of original case studies. The philosophies of particular sciences are both drawing on and feeding into new work in metaphysics and the relationships between science, metaphysics and the philosophy of science in general are being re-examined and reconfigured. The intention behind this new series from Palgrave-Macmillan is to offer a new, dedicated, publishing forum for the kind of exciting new work in the philosophy of science that embraces novel directions and fresh perspectives. To this end, our aim is to publish books that address issues in the philosophy of science in the light of these new developments, including those that attempt to initiate a dialogue between various perspectives, offer constructive and insightful critiques, or bring new areas of science under philosophical scrutiny. The members of the editorial board of this series are: Otavio Bueno, Philosophy, University of Miami (USA) Anjan Chakravartty, University of Notre Dame (USA) Hasok Chang, History and Philosophy of Science, Cambridge (UK) Steven French, Philosophy, University of Leeds (UK) series editor Roman Frigg, Philosophy, LSE (UK) James Ladyman, Philosophy, University of Bristol (UK) Michela Massimi, Science and Technology Studies, UCL (UK) Sandra Mitchell, History and Philosophy of Science, University of Pittsburgh (USA) Stathis Psillos, Philosophy and History of Science, University of Athens (Greece) Forthcoming titles include: Sorin Bangu, Mathematics in Science: A Philosophical Perspective Gabriele Contessa, Scientific Models and Representation Michael Shaffer, Counterfactuals and Scientific Realism Initial proposals (of no more than 1000 words) can be sent to Steven French at s.r.d.french@leeds.ac.uk.

More information about this series at
http://www.springer.com/series/14743

Lena C. Zuchowski

A Philosophical Analysis of Chaos Theory

Lena C. Zuchowski
Fachbereich Philosophie
Universität Salzburg
Salzburg, Austria

New Directions in the Philosophy of Science
ISBN 978-3-319-85448-9   ISBN 978-3-319-54663-6 (eBook)
DOI 10.1007/978-3-319-54663-6

Cover illustration: Détail de la Tour Eiffel © nemesis2207/Fotolia.co.uk

Printed on acid-free paper

This Palgrave Macmillan imprint is published by Springer Nature
The registered company is Springer International Publishing AG
The registered company address is: Gewerbestrasse 11, 6330 Cham, Switzerland

ACKNOWLEDGEMENTS

I am very grateful to Jeremy Butterfield and Charlotte Werndl for their help with this project. I also thank Maddie Geddes-Barton and Jack Blaiklock for their help with the manuscript.

CONTENTS

LIST OF FIGURES

LIST OF TABLES

CHAPTER 1

Introduction

Abstract The book's strategy and content will be introduced. Over the last 40 years, chaos theory has had a huge impact on science and philosophy. This is evidenced by the astonishing volume of chaos-related publications; even a cursory survey shows that chaos has been detected virtually everywhere, from cardiac rhythms to Joyce's Ulysses (e.g. Kellert, 2008). Given the high export appeal of chaos theory, it is surprising that there are fundamental aspects of the field that still remain poorly understood and, in some cases, permanently debated.

Keywords Chaos · Introduction

In particular: (i) it is still not clear how chaos should be defined and how the large number of coexisting chaos definitions relate to each other (e.g. Smith, 1998; Werndl, 2009c); (ii) there are still (largely unarticulated) questions about the faithfulness and predictiveness of the numerical and theoretical models on which chaos theory is based; and, finally, (iii) it has not been unequivocally resolved whether there is chaos in nature (e.g. Kellert, 1993) and how it should be diagnosed (e.g. Pool, 1989; Hastings et al., 1993). The three aspects are not independent of each other and it is evident that difficulties (i) and (ii) contribute to difficulty (iii). Together, they have made it very difficult to judge the diverse contributions to chaos theory comparatively and to enforce universal standards of quality and rigour.

1

L.C. Zuchowski, *A Philosophical Analysis of Chaos Theory*, New Directions in the Philosophy of Science, DOI 10.1007/978-3-319-54663-6_1

This book aims to clarify aspects (i)–(iii) by providing a structured survey of the construction, diagnosis and evaluation of chaotic models. Although the book follows a survey approach in that it aims to achieve a certain degree of comprehensiveness and to comparatively cover different aspects of chaos theory, it is not a mere review. I will pursue a modelling-centred strategy and thereby aim to provide the first in-depth analysis of all three stages, i.e. construction, diagnosis and evaluation, of modelling in chaos theory. This allows me to draw on a large amount of recently developed work on the use of models in science, which has so far not been applied to the field of chaos theory. In particular, the book uses, and develops further, several results of both the fictionalist approach to modelling (e.g. Frigg, 2010; Toon, 2012; Suarez, 2013) as well as the work on horizontal modelling by Bokulich (2003), which were not available to authors of earlier philosophical analyses of chaos theory (e.g. Kellert, 1993; Smith, 1998).

Throughout the book, I will follow a strategy of rational reconstruction, that is I will aim to develop analytic frameworks that allow the exposition of the relevant concepts with the greatest possible clarity. These frameworks are intended to be tools in the conceptual analyses presented here. I thereby neither aim to be descriptively accurate on an individual level, that is to paint a detailed picture of the individual practices of chaos scientists, nor normatively prescribing, that is to offer advice to scientists on how the reconstructed concepts should be used. Instead, I aim to develop frameworks in which the content and the use of a given concept can be displayed clearly. The merits of the analytic frameworks developed in this book should therefore be evidenced in the quality of the analyses provided.

The book is subdivided into three main content chapters. These chapters deal with the construction, diagnosis and evaluation of chaotic models, respectively.

In Chapter 2, I will analyse the construction of vertical and horizontal models in chaos theory. I will review some selected material on scientific modelling, which will be crucial to the discussion in the remainder of the book (Section 2.2). In particular, I will discuss the construction and evaluation of two different classes of models: vertical and horizontal models. I will introduce the inferential account of model evaluation developed by Suarez (2013). This account conceptualizes the evaluation of vertical models as the evaluation of a conditional $C \rightarrow B$ to be transferred from the model to the target system. Underlying this framework is the assumption that scientists are not only interested in the occurrence of a certain behaviour B but also in the sufficient conditions C for this behaviour. Within this framework for the analysis of model evaluation, the concept of model faithfulness will be defined.

I will then present two case studies to illustrate the construction of vertical and horizontal models in chaos theory: the construction of three models based on the logistic equation (Section 2.3) and of two models based on the Lorenz equations (Section 2.4).

In Chapter 3, I will analyse the criteria for and definitions of chaos. The chapter has two main theses: (i) I will maintain that there are five main criteria to diagnose chaos and (ii) that various combinations and embodiments of these criteria are used to build different chaos definitions.

I will begin this chapter by introducing an analytic framework that views chaos definitions as twofold decomposable: into five main criteria and into different technical embodiments of these criteria. The use of this framework will then be illustrated in a case study of the diagnoses of chaos of the logistic models (Section 3.2). This case study will introduce the five criteria that I consider to be constitutive of virtually all existing chaos definitions: determinism, transitivity, periodicity, aperiodicity and sensitive dependence on initial conditions (SDIC).

I will then discuss these five criteria in detail (Section 3.3). I will show that the criteria are similarity categories and can assume many different formal or semi-formal embodiments. It will become apparent that the different embodiments of a criterion can be used to make this criterion applicable to a specific class of models. The fact that many embodiments are applicable only to a specific class of models mitigates any conceptual conflicts between the different criteria.

I will then demonstrate how the most prevalent chaos definitions are composed of different combinations of embodiments of the five core criteria (Section 3.4). Five definitions will be analysed in detail: Devaney chaos; the definition of chaos as mixing; the definition of chaos in terms of positive Lyapunov exponents; stochastic chaos; and the definition of chaos in terms of strange attractors. I will maintain that many of these definitions are targeted towards specific classes of models and that the use of different combinations of criteria in different definitions can be viewed as a means of highlighting those properties of these models that will be the most important for their future investigative use. The coexistence of many different chaos definitions can therefore be viewed as a consequence of the variety of models used in chaos theory.

In Chapter 4, I will discuss the evaluation of models in chaos theory. Building on the framework for the transference of conditionals from models to their target systems developed by Suarez (2013), I will show that, for vertical chaotic models, the evaluation process can be decomposed into three crucial steps: (i) one needs to determine which type of chaos and which proposed sufficient conditions for its occurrence are to be evaluated, that is one needs to determine a conditional C → B to be transferred from the model to the target system; (ii) the existence of chaos in the model's target system needs to be ascertained and (iii) the faithfulness of the model should be evaluated. While it is possible to clearly separate these steps conceptually, actual evaluations of vertical chaotic models do not necessarily employ these steps in order or give equal weight to all steps. The use of this analytic framework will be demonstrated in a case study of the evaluation of the logistic models (Section 4.2).

I will then discuss each of the three steps in detail (Section 4.3). It will become apparent that there are two types of chaotic conditionals to be transferred from vertical models: conditionals that posit forms non-linearity and iteration as sufficient conditions for the occurrence of chaos (type 1); and conditionals that posit non-linearity and discreteness as sufficient conditions for chaos (type 2). It will be shown that the determination of which conditionals holds true in a model is usually technically difficult and often involves investigative work with related horizontal models. I will therefore be able to specify the investigative function of horizontal models more precisely as aiding investigations of the properties of related vertical models during the first step of these latter models' evaluation. The use of horizontal models in chaos theory will also be discussed in a separate section of the chapter (Section 4.4).

I will maintain that the second step of the model evaluation process, that is determining the existence of chaos in the target system, is often difficult to complete for chaotic models. This difficulty will be traced back to a difficulty in determining the fulfilment of the criterion of determinism.

Finally, I will discuss the model faithfulness of chaotic models. A particular conceptual result of this analysis will be the realization that numerical models can only model chaos faithfully, if the chaotic behaviour is seen as part of a type 1 conditional. Accordingly, a significant part of the modelling activity of scientists in chaos theory can be interpreted as attempts to establish these conditionals in numerical models. This last point, and the general interplay of different models in chaos theory, will then be illustrated in a case study of the evaluation of the Lorenz models (Section 4.5).

CHAPTER 2

Vertical and Horizontal Models in Chaos Theory

Abstract I will review material on vertical and horizontal modelling. The vertical and horizontal construction of models in chaos theory will be demonstrated in two case studies: the construction of three models based on the logistic equation and two models based on the Lorenz equations.

Keywords chaos · logistic model · Lorenz model · models in science

2.1 INTRODUCTION

In the first section of this chapter, I will introduce some selected material on scientific modelling. In particular, I will discuss the construction and evaluation of two different classes of models: vertical (Section 2.2.1) and horizontal (Section 2.2.2) models. During the discussion of vertical models, a general framework for the rational reconstruction of the process of model evaluation based on transferences of conditionals between a model and its target system will be introduced.

I will then present two case studies to illustrate the construction of vertical and horizontal models in chaos theory. In Section 2.3, I will describe the construction and behaviour of three models based on the logistic equation: the continuous logistic model (Section 2.3.1) and the discrete logistic model (Section 2.3.2), which have both been constructed vertically, as well as the iterated logistic model, which has been constructed horizontally (Section 2.3.3).

L.C. Zuchowski, *A Philosophical Analysis of Chaos Theory*, New Directions in the Philosophy of Science, DOI 10.1007/978-3-319-54663-6_2

In Section 2.4, I will describe the construction and behaviour of two models based on the Lorenz equations: the well-known discrete model (Section 2.4.1), developed by Lorenz (1963), and an iterated model (Section 2.4.2), which is constructed horizontally from the discrete Lorenz model.

To my knowledge, very few detailed studies of the construction of models in chaos theory have been conducted. The few existing studies that provided more detailed analyses of the development of chaotic models (Kellert et al., 1990; Koperski, 1998) deal exclusively with vertical models and focus on the evaluation of these models. My work in this chapter is therefore novel in two respects: (i) in focusing on model construction rather than on model evaluation (which will be addressed in Chapter 4); and (ii) in investigating the use and importance of horizontal models in chaos theory. My analysis will draw on a large amount of recent philosophical work on the use of models in science, which has so far not been applied to the field of chaos theory. In particular, I will use and develop further several results from both the fictionalist approach to modelling (e.g. Frigg, 2010; Toon, 2012; Suarez, 2013) as well as the work on horizontal modelling by Bokulich (2003), which were not available to authors of earlier philosophical analyses of chaos theory (e.g. Kellert, 1993; Smith, 1998).

Four of the models discussed here have achieved iconic status in chaos theory; and their influences on the different definitions of chaos will be outlined in Chapter 3. The five models introduced here will be used as case studies throughout the book.

2.2   Vertical and Horizontal Models

Models and their uses in science have been the subject of a prolific and ongoing debate in the philosophy of science (e.g., for review, Frigg and Hartmann, 2012). In fact, there is so much material available on the various aspects of scientific modelling that keeping the review portion of this book to appropriate proportions meant disregarding much of it.

My selection of material has been guided by the needs of my particular study and the unique situation of chaos theory as a scientific field. In particular, chaos theory is an interdisciplinary field, located at the intersection between the natural sciences, pure mathematics and the computational sciences. However, much of the philosophical debate on modelling focuses exclusively on scientific modelling as practiced by natural scientists. According to Bokulich (2003, p. 610), the prototypical model ascribed to these scientists is a "vertical" one: a model that has been constructed "either top-down from theory or bottom up from empirical data". Philosophical analyses have then, naturally, been focused on exploring the relationships between these different vertically organized levels. Standard philosophical questions in this context are: (i) how models have been constructed from the appropriate top-level theory; and (ii) how well they represent their bottom-level, empirical target systems.

Many chaotic models are vertical models and I will therefore briefly review the necessary material on vertical modelling in Section 2.2.1. My review will introduce two key concepts for my later analysis of the construction and evaluation of vertical chaotic models: a conceptualization of model evaluation that is based on the transference of conditionals (Suarez, 2013) and a notion of model faithfulness that is compatible with this interferential framework.

However, I also maintain that there exists a second class of models in chaos theory, which so far has not received much attention from philosophers. Bokulich (2003, p. 611) describes these as "horizontal models": models that are not constructed vertically from theory or empirical data but are horizontal spin-offs from existing models. The general construction and evaluation of horizontal models will be described in Section 2.2.2.

To my knowledge, little further work on horizontal models has been under-taken since Bokulich's (2003) original study. In my opinion, the notion of horizontal modelling could be helpful to explain the use of models in other interdisciplinary fields, for example, complexity science and some parts of economics, as well; and I am hoping that my work here will also add to the general exploration of this "neglected" class of models.

2.2.1   Vertical Models

Toon (2012, p. 9) provides the following description of modelling, which I consider to be a good example of the definitions customarily used by philosophers:

> [However,] I shall use 'theoretical modelling' in a broader sense, to include any case in which scientists deliberately simplify or idealize a system in order to explain or predict its behaviour . . .

Examples of theoretical modelling listed by Toon (2012, pp. 7–9) include: modelling a bouncing spring as a harmonic oscillator; modelling ideal gases as billiard balls; the construction of the Lotka-Volterra model of predator–prey interaction; the construction of general equilibrium models of markets and several additional examples. The term "theoretical" here is meant to exclude "physical", that is material, models from the definition. Since there are relatively few physical models in chaos theory, I too will only be concerned with theoretical models and will therefore drop the adjective in the following.

In the following paragraphs, I will outline the construction and evaluation of vertical models, using Toon's (2012, p. 9) definition as a starting point for this discussion.

Construction of Vertical Models

In Toon's (2012) definition, the relationship between the target system and the model is described as "simplification" and "idealization". Toon (2012, p. 10) describes these as instances of what Cartwright (1983, p. 15) has named "prepared descriptions": specifications of a number of alterations to be made to a set of general equations to fit a particular, empirically determined scenario. Thereby, the set of general equations is derived from the top-level theory, which prima facie includes all natural laws governing the phenomenon under investigation. The process of paring down the full governing theory to the smaller set of equations comprising the model is seldom fully articulated; rather, scientists seem to pre-select a small number of relevant governing laws and then to prescribe additional modifications to these. For example (Toon, 2012, pp. 7–9), in the construction of the classic harmonic oscillator, the equation selected is Newton's second law together with a prepared description specifying the removal of all terms and spatial integrals that derive from frictional and inertial forces. The result of applying the prepared description to Newton's second law is the well-known harmonic equation.

I prefer the term "prepared description" over "simplification and idealization" (see quotation above) since the former can include any technical alteration to the general equations without having the connotation of providing an "easier" description. In particular, in chaos theory, prepared descriptions often include the process of discretization (e.g. Section 2.3.2 and Section 2.4.1). In such cases, it is far less obvious that discretization really provides a simpler description of a phenomenon than it is to claim, say, that neglecting friction provides a simpler model of the harmonic oscillator.

As described earlier, in my opinion, the construction of vertical models usually involves both the governing theory as well as the empirical target system. I therefore differ from Bokulich's (2003, p. 610) opinion that there are two possible (exclusive) ways of vertically constructing a model: top-down from theory or bottom-up from empirical data. I agree that there exist cases of pure bottom-up construction, for example those in which a function is fitted mathematically to existing empirical data. In such cases of pure curve-fitting, no knowledge of the governing theory is necessary and the model provides a purely phenomenological description of the target system.

Koperski (1998) claims that the bottom-up construction of models is the canonical form of model construction in chaos theory. I disagree with this claim and, in Section 4.3.2, will argue that the activity (Koperski, 1998) describes as bottom-up modelling – phase-space reconstruction – should rather be viewed as a tool to diagnose chaos in natural systems. However, even those cases of model construction that involve top-level theory, that is in cases of top-down construction, empirical information about the target system is used for the construction in the form of the prepared description. This view of the construction of vertical models implies that the target system cannot be used as an independent empirical means to evaluate the model against. As outlined in the next paragraph, the evaluation of vertical models is therefore not straightforward.

Evaluation of Vertical Models

The question of how vertical models can and should be evaluated against their target systems has generated much philosophical discussion. Fundamental philosophical worries about such evaluations primarily stem from the fact that the model has been constructed from a prepared description of the target system: it is therefore a representation of the target system, which is known – and intended – to differ from the system itself. However, the major epistemic role ascribed to vertical models is being informative about their target system (e.g. Bailer-Jones, 2002; Bolinska, 2013). The evaluation of a model therefore entails judging how trustworthy information gained from the model are in guiding our knowledge of, and expectations about, the target system.

Here, I will only address two philosophical issues in the evaluation of vertical models: (i) the semantic difficulty of comparing the abstract model to the concrete target system; and (ii) the transformation of predictions (roughly cast in the form of a logical conditional) made in the model into expectations about the target system.

With respect to issue (i), there exists the semantic difficulty that the model, as constructed from top-level theory and the prepared description, consists only of a set of equations. Statements about the model therefore refer to an abstract object, while statements about the target system refer to a concrete system. Giere (1988) solves this difficulty by introducing the concept of "model system": that is the hypothetical concrete system that would be described exactly by the model's equations. Statements about the model system are then formally comparable to those about the target system. While Giere's (1988) solution has been criticized (e.g. Levy, 2015), I will follow the majority of philosophers in adopting this approach and in describing the evaluation of vertical models in terms of comparisons between model systems and target systems.

Even after this semantic difficulty has been resolved, comparisons between a model and a target system still involve the somewhat paradoxical situation of expecting similarities between two systems of which one has been designed to differ from the other. A recent, influential approach to this problem interprets the relationship between model and target systems to be analogous to that between reality and literary fiction (e.g. Godfrey-Smith, 2009; Contessa, 2010; Frigg, 2010; Toon, 2012). Under this interpretation, model systems can be viewed as entities that are known to contain false, made-up parts but can still be compared to real systems in a meaningful way. It is thereby important that knowledge of the entity's fictional nature is shared: for example it is permissible to say that a friend possesses deductive powers comparable to that of Sherlock Holmes only if the maker of the statement (and all its interpreters) understand that Sherlock Holmes is fictional while the friend is not. Similarly, a comparison between a model system and a target system would go awry, if the scientist was not aware of the constructed nature of the former. These similarities in the epistemic natures of fictions and models have prompted descriptions of modelling as engaging in "authorized games of make-believe" (Frigg, 2010, p. 259).

In the fictionalist account of modelling, the fictional parts of a model can be viewed as encoded in the prepared description, which might ask us to imagine, say, a world in which there exists no friction but which contains point masses. Other parts of the prepared description might be judged as non-fictional if checked against the target system, for example, statements to the effect that a particular harmonic oscillator involves a spring and that its spring constant has a certain value k. It is therefore usually possible to at least roughly identify where the model system, as described by the model's equations, differs from the target system. However, it is much more difficult to evaluate how new information gained from the model system transfers to the target system. For example, if the model system reliably displays a specific behaviour under certain conditions, should scientists then expect to find the same behaviour under comparable conditions in the target system? Potential answers to this question will be discussed in the next paragraph.

Model Faithfulness and the Transference of Conditionals
Bolinska (2013, p. 229) links the notion of information transfer between model and target systems to the concept of representational faithfulness:

> An aspect of a target system is faithfully represented to the extent that a user is able to draw true inferences about that aspect from the vehicle [model].

Bolinska's (2013) concept of faithfulness is both "a notion that admits of degrees" (p. 230) and one that is retrospectively assigned: only after the inferences have been drawn and compared to the actual behaviour of the target system can one assign a fractional value of faithfulness. An example used by Bolinska (2013) is that of different maps of a city: a recently updated version of the map is usually found to provide a more faithful representation than an older version.

For scientific models, and, in particular, those used in chaos theory, evaluations of model faithfulness often prove to be more difficult. Suarez (2013) discusses the evaluation of predictions of the form C → B, that is statements of the form "if conditions C exist, then behaviour B will be displayed" when derived from models that have known fictional parts. Suarez (2013) is interested in the question under which circumstances scientists would be willing to accept the conditional C → B as true in the target system, that is under which circumstances scientists would be willing to accept C as a sufficient condition for B (as apparent in the model system), while knowing that parts of the model system are fictional.

Before considering the transference of the conditional to a target system, it is important to clarify what it means to say that a conditional C → B is true in a model. Formally, the conditional states that whenever C is present, B must also be present. However, the reverse is not true: the behaviour B can be displayed while the condition C is not given. Accordingly, the conditional only establishes a weak form of causation by requiring that C is a sufficient (but not a necessary) condition for B. However, holding the conditional true entails that C is seen as predictive of B. Therefore, whenever the condition C has been observed (or engineered to exist) in the model, then the behaviour B should be displayed. In this framework, the evaluation of the model with respect to a conditional C → B therefore involves determining whether the condition C should also be seen as sufficient for the behaviour B in the target system. As I will discuss in Section 4.3.1, determining which conditionals C → B hold true in a chaotic model usually requires significant investigative work.

As a precondition for any consideration of the transference of a conditional C → B, Suarez (2013) assumes that the behaviour B is observed in the target system. In Section 4.3.2, I will show that establishing that the target system really shows a specific kind of chaos, that is determining that the behaviour B is really observed in the target system, can itself be difficult. The fact that much effort is expended by scientists on determining that chaos exists in the natural systems targeted by chaotic models is a peculiarity of chaos theory as a scientific field. It also provides support for Suarez's (2013) view of this step as crucial to the evaluation process.

Suarez (2013, p. 246) assumes that the conditional C → B will be expected to transfer to the target system, either (i) if the fictional parts of the model system deriving from the prepared description are not part of the antecedent C, but can be relegated to another set of background conditions C_b, that is if the conditional properly reads C_b → (C → B); or (ii) if this is not the case but non-standard semantics apply, that is if the conditional C → B can be assigned an acceptable truth-value even if C is true and B is false. I will subsequently focus on the former condition since there is no indication that non-standard semantics govern the acceptance of such conditionals by scientists in chaos theory.

Why would scientists accept the transference of the conditional if no fictional parts appear in the condition C? The formal consequence of relegating all fictional parts into C_b is that C_b will be false and C_b → (C → B) will be true, regardless of the value of C → B. Therefore, the truth value evaluation of C → B has formally been decoupled from that of the background conditions C_b. Conceptually, one can view the possibility of separating the background assumptions C_b from the condition C as a means to ensure that the condition C is independent of the particular set-up of the model system and that it could hence be precisely reproduced in the target system. Since the core conditional C → B now only depends on factors that occur in both the model and the target system, inferences from this conditional carry across to the target system. In other words, one needs to ensure that only aspects that are germane to both the model and the target system are sufficient conditions for the occurrence of the behaviour one wishes to predict or explain in the target system.

Suarez (2013) sees a major advantage of his account of the evaluation of models in its compatibility with both anti-realist as well as realist philosophies of science. For my own analysis of modelling in chaos theory, I see an additional advantage: the framework of model evaluation based on the transference of conditionals $C \to B$ allows for a clear separation between the conditions under which a certain behaviour is expected to occur in a model, or a natural system, and the definition of this behaviour itself. Both the question of what is predictive of chaos in a model or a natural system and the question of how chaos should be defined are still contested. By providing an analysis that differentiates sharply between sufficient conditions for the occurrence of a behaviour and the criteria used to diagnose this behaviour, I hope to address both questions more clearly. This distinction has not been focussed on in previous philosophical accounts of chaos theory (e.g. Smith, 1998; Schurz, 1996; Kellert, 1993). I therefore hope that my work here will add to the existing literature by using the natural separation of sufficient conditions C and behaviour B inherent to Suarez' (2013) framework as a structuring device.

So, combining the two concepts gleaned from Bolinska (2013) and Suarez (2013), I propose a more specific notion of model faithfulness:

Definition 1. An aspect B of a target system is faithfully represented in a model, if it is conditionally isolated in the model system from all fictional aspect arising from the construction of the model, that is, if an appropriate conditional $C \to B$ can be expected to transfer from the model to the target system.

This notion of model faithfulness, and the methodological framework based on the transference of conditionals outlined above, will provide the framework for my discussion, in Section 4.3, of the evaluation of vertical chaotic models. In particular, I will show that this process of evaluation can be further decomposed conceptually into three distinct steps: (i) the determination of the conditional $C \to B$ to be transferred from the model system to the target system (Section 4.3.1); (ii) the determination of the existence of the behaviour B in the target system (Section 4.3.2) and (iii) the determination of model faithfulness with respect to the behaviour B (Section 4.3.3). Furthermore, I will be able to further specify the investigative role of horizontal models in chaos theory by showing that, during step (i), they are used as tools to investigate the properties of a related vertical model and to determine the conditional C → B to be transferred (Section 4.4).

2.2.2   Horizontal Models

As foreshadowed in the introduction to this chapter, I agree with Bokulich (2003) that there exists a second class of models besides the vertically constructed models discussed in Section 2.2.1: namely, horizontally constructed models. These models differ from vertical models in that they are not constructed from top-level theory and bottom-level empirical data but are developed as "free" variations of existing models. Accordingly, both the concept of model faithfulness (Definition 1) and the last two steps of the framework for the evaluation of horizontal models are not applicable to horizontal models: these models need not be evaluated against a target system since no such target systems exist.

I will show throughout this book that many chaotic models are horizontal models (e.g. Section 2.3.3 and Section 2.4.2) and that the interaction between these two classes of models is one of the most characteristic features of chaos theory as a scientific field (Chapter 4). In this section, I will briefly discuss some details on the general construction and evaluation of horizontal models. Other than Bokulich (2003), there seem to be virtually no studies focussing explicitly on horizontal modelling: this review will therefore heavily rely on Bokulich's (2003) study.

Construction of Horizontal Models
Bokulich (2003, pp. 612) begins her study with a description of how models in the field of semi-classical physics are constructed: by discretizing the equations of a particular classical model. Thereby, the question of how the original model itself was constructed is unimportant; no information about its governing theory and its target system needs to be transmitted. The initial model merely provides a set of equations, which are mathematically manipulated to provide the new horizontal model. Through this process of construction horizontal models become "part of a lineage of models with their own internal dynamics and justification" (Bokulich, 2003, p. 613). In my case studies of the logistic models (Section 2.3) and the Lorenz models (Section 2.4), I will show that such lineages can clearly be identified in chaos theory.

Evaluation of Horizontal Models

Since horizontal models are not constructed from empirical data and governing theories, their epistemic function is different from that of vertical models (Section 2.2.1). In particular, neither can their primary epistemic function be to gain information about a target system, nor can they be evaluated against such a system. Instead, Bokulich (2003, p. 613) assigns horizontal models an investigative function: for example, in the case of semi-classical physics, they are a means of investigating relationships between classical and quantum theories.

I agree with Bokulich (2003) that horizontal models have an investigative function rather than one of mediating between theory and data. However, I take this function to be both more general, as well as more subject-specific, than that of investigating inter-theoretical relationships. Horizontal models in chaos theory often appear to be designed with the explicit aim of gaining more information about specific other chaotic models; usually with the aim of establishing the sufficient conditions under which this model will behave chaotically (e.g. Sections 4.2.2 and 4.5). Using the framework for the evaluation of vertical models developed in Section 4.3, I can therefore locate the investigative use of horizontal models in the first conceptual step of the evaluation of a related vertical model, that is, it can be viewed as part of the determination of the conditional $C \rightarrow B$ to be transferred from such a related model. The origin of the generating model can thereby be of even less importance than in Bokulich's (2003) account: the fact that its generating model might have had importance in a certain branch of science is usually accidental rather than crucial to the construction of a horizontal model.

In Section 4.4, I will outline in more detail how the construction of horizontal models from vertical ones appears to be one of the main mechanism that make chaos theory an interdisciplinary field. However, horizontal models in chaos theory do not appear to be primarily used in the investigation of inter-theoretical relationships (as found by Bokulich [2003] in her case study of semi-classical physics). Instead, these models are a means of applying abstract results from the field of mathematics to models used in the applied sciences (e.g. biology and meteorology, Section 2.3 and 2.4).

Horizontal models in chaos theory therefore constitute interdisciplinary "bridges" but their investigative role is not that of exploring inter-theoretical relationships.I suspect that the investigative role of horizontal models is highly subject-specific and that more research on these models in other fields of science is needed before a comprehensive understanding of their epistemic functions can be achieved.

The evaluation of horizontal models therefore consists in the comparison with, and the investigation of, other models in their lineage. In contrast to vertical models, horizontal models are not used directly to gain information about real-world systems. However, they are used as tools in the evaluation of vertical models and therefore indirectly contribute to the exploration of those models' target systems. The transference of results from a horizontal model to a related vertical model can be difficult, in particular, for lineages of dynamically complex models. In Section 4.5, I will illustrate such difficulties with a case study of Smale's 14th problem, that is the question of whether the vertically constructed discrete Lorenz model is chaotic in the same sense as the horizontally constructed iterated Lorenz model.

2.3   A Lineage of Logistic Models

In this section, I will discuss the construction of three models associated with the logistic equation. This section, and the following Section 2.4, serves two functions within this book: firstly, they illustrate the processes of model construction discussed in Section 2.2.1 and Section 2.2.2; secondly, they introduce four iconic chaotic models, which will serve as case studies throughout this book.

The name "logistic equation" is given to the following non-linear, first-order differential equation (e.g. Hilborn, 2002; Devaney, 1989):

$$\frac{dN}{dx} = rN(1 - N) \tag{2.1}$$

where N is the dependent variable, x is the independent variable and r is a constant.

2.3.1   Continuous Logistic Model

May (1974, p. 645) introduces eq. (2.1) as part of the construction of a model of population dynamics:

> In some biological populations (for example, man), growth is a continuous process and generations overlap; the appropriate mathematical description involves non-linear differential equations.

Construction of the Continuous Logistic Model
Equation (2.1) is named as one of "the simplest such differential equations", where N is interpreted as population number, r is a growth rate and x a continuous time parameter. May (1974, p. 645) explicitly points out that the model is simplified because it is assumed that the change in population number N depends on its immediate value; it would be more true to the target system to assume a time delay to the change in population growth. He does not discuss the fact that approximating overlapping generations with a continuous time variable x is also an approximation but this is, of course, implicit in the introductory quotation. Clearly, the continuous model of the logistic equation is therefore a vertical model (Section 2.2.1), constructed from a general set of laws governing population dynamics and a prepared description of a specific class of such populations.

Equation (2.1) can be analytically integrated so that the model can be displayed in time-explicit form:

$$N(x) = \frac{N_0 e^{rx}}{1 + N_0(e^{rx} - 1)} \tag{2.2}$$

where N_0 is an initial value for $N(x)$.

Behaviour of the Continuous Logistic Model
It is immediately apparent that, independent of the initial value N_0, for $x \rightarrow \infty$, $N(x)$ monotonically approaches $N(x) = 1$ for all positive values of r and monotonically approaches zero for all negative values of r. The dynamics of the continuous logistic model system is hence one where the population number eventually approaches an equilibrium value. This model of the logistic equation shows monotonic behaviour and is not chaotic (Section 3.2.1).

2.3.2   Discrete Logistic Model

May (1974, p. 645) then introduces a different vertical model based on eq. (2.1), which is tied to the following target system:

> In other biological situations (for example, in 13-year periodical cicadas), population growth takes place at discrete intervals of time and generations are completely non-overlapping; the appropriate mathematical description is in terms of nonlinear difference equations.

Construction of the Discrete Logistic Model

Equation (2.1) can be prepared to fit this situation by discretization:

$$dN = rN(1 - N)dx,$$

$$N_{x+1} - N_x = rN_x(1 - N_x)\Delta x \qquad (2.3)$$

where the subscript x denotes the xth iteration in time and Δx is the difference between two such steps. In the case of non-overlapping generations, one can assume $\Delta x = 1$ and hence:

$$N_{x+1} = N_x(1 + r(1 - N_x)) \qquad (2.4)$$

It is notable that x here is a discrete variable rather than a continuous one as in Section 2.3.1. The values obtained from eq. (2.4) for different values of the growth rate r are plotted in Fig. 2.1.

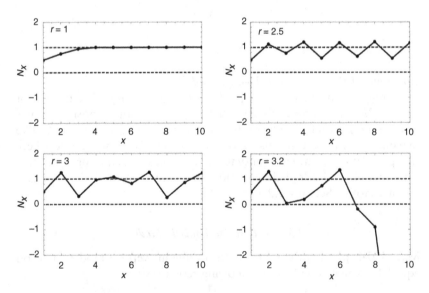

Fig. 2.1   Behaviour of the discrete logistic model. The plot shows the population values N_x (dots). The model uses an initial value of $N_0 = 1/2$. The unit interval in N_x spaces is indicated by dashed lines

Behaviour of the Discrete Logistic Model

The behaviour of the discrete logistic model (2.4), together with a number of similar vertical population models, has been studied in detail by May and Osler (1976). My discussion will roughly follow theirs. However, I do not aim to present a formal mathematical analysis; and, whenever possible, I will illustrate properties by referring to Fig. 2.1 rather than proving their existence formally.

The development of N_x for growth rates below $r = 2$ is illustrated in the first panel of Fig. 2.1. The map N_x will converge to $N_x = 1$, which eventually becomes a fixed point. The speed of this convergence is proportional to the growth rate r. The behaviour of the discrete model (2.4) therefore closely mirrors that of the continuous model (Section 2.3.1).

For growth rates between $r = 2.449$ and $r = 2.57$, the behaviour of eq. (2.4) is periodic, that is all N_i will repeat themselves at N_{i+p}, $N_{i+2p} \ldots$, where p is some period of repetition. This parameter regime is also illustrated in Fig. 2.1.

About the behaviour of the discrete logistic model beyond $r = 2.57$, May and Osler (1976, p. 583) write:

> [T]here are an infinite number of periodic points.... Furthermore, there are an uncountable number of points (initial conditions) whose trajectories are totally aperiodic; no matter how long the time series generated by [the difference equation] is run out, the pattern never repeats [...].

The aperiodic behaviour of the model is illustrated in the two lower panels of Fig. 2.1.

A fourth regime is identified by May and Osler (1976, p. 586–588): for growth rates larger than $r = 3$, the discrete population values N_x will eventually decrease monotonically. This is described as "extinction". The last panel of Fig. 2.1 illustrates the behaviour of the model in this regime.

The behaviour of the discrete model system is therefore one where non-overlapping populations develop in various ways depending on the growth rate r. This development is very different from that of the continuous model system (Section 2.3.1). In Section 3.2.2, I will show that May and Osler (1976) find the model to be stochastically chaotic; in Section 3.4 we will see that it is also chaotic according to several other definitions of chaos. The continuous logistic model, although constructed from the same general theory, is not chaotic. This indicates that the differences in the models' prepared descriptions, that is the presence or absence of discretization, are among the sufficient conditions for the different behaviours displayed. Further evidence for this is provided through investigative work with a related horizontal model, the iterated logistic model, which I will introduce in Section 2.3.3. Since the continuous and the discrete logistic models have different target systems, this is *prima facie* not worrisome. In Sections 4.2.1 and 4.2.2, it will also be shown that discretization should not be viewed as a fictional part of the model as developed by May and Osler (1976). Therefore, the various kinds of chaotic behaviour displayed in the discrete logistic model can be assumed to be modelled faithfully (in the sense of Definition 1, Section 2.2.1).

2.3.3   Iterated Logistic Model

My description of the construction and behaviour of the iterated logistic model will be based on the discussion by Devaney (1989). Historically, the model seems to have been first introduced in two papers aiming to provide further discussion of the results of the discrete logistic model developed by May (1974), Li and Yorke (1975) and Guckenheimer et al. (1977). The precise relationship between the two models will be discussed in Section 4.2. Since Devaney (1989) provides a more comprehensive review of the construction and behaviour of the iterated logistic model, my discussion here will be based on this account. As in Section 2.3.2, I do not aim to provide formal justifications or mathematical comprehensiveness: instead, the reader is referred to the original sources to look up formal proofs and derivations. Whenever possible, I will refer to Figs. 2.2 and 2.3 to illustrate the claims made in this section.

Construction of the Iterated Logistic Model
Devaney (1989, pp. 5–7) constructs the iterated model of the logistic equation by successive self-application of the right-hand side of eq. (2.1) to itself:

$$N^{(0)}(x) = x,$$

$$N^{(1)}(x) = rx(1 - x),$$

$$N^{(2)}(x) = rN^{(1)}(x)\left(1 - N^{(1)}(x)\right),$$

$$\dots$$

$$N^{(n+1)}(x) = rN^{(n)}(x)\left(1 - N^{(n)}(x)\right). \qquad (2.5)$$

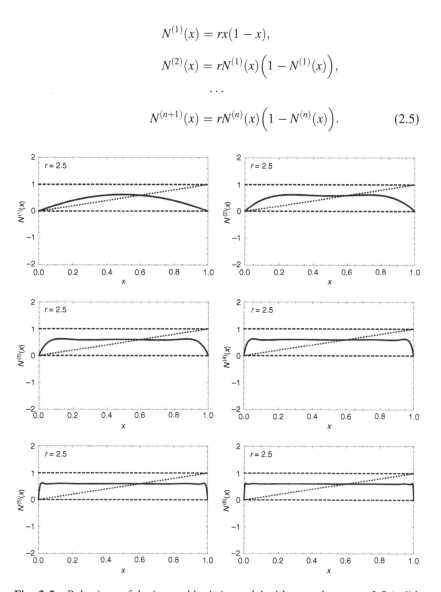

Fig. 2.2 Behaviour of the iterated logistic model with growth rate $r = 2.5$ (solid line) in comparison to the identity function (dotted line). The unit interval is indicated by dashed lines

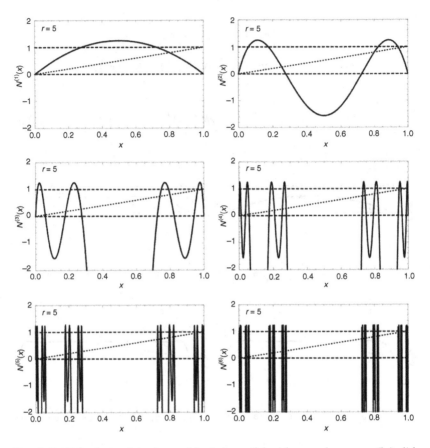

Fig. 2.3 Behaviour of the iterated logistic model with growth rate $r = 5$ (solid line) in comparison to the identity function (dotted line). The unit interval is indicated by dashed lines

It is easy to see that this will result in consecutively higher polynomial terms in x:

$$N^{(0)}(x) = x,$$
$$N^{(1)}(x) = rx - rx^2,$$

$$N^{(2)}(x) = r^2 x - (r^2 + r^3)x^2 + 2r^3 x^3 - r^3 x^4,$$

$$\cdots$$

The order of the polynomial in x will be n^n for even n and $(n-1)^{2(n-1)}$ for odd n. The order of the polynomial in r will be $(n-1)^{n-1} + n$.

In contrast to the continuous (Section 2.3.1) and the discrete (Section 2.3.2) logistic model, neither the iteration variable n nor the independent variable x in eq. (2.5) is necessarily time-like. Devaney (1989) also does not tie his construction of the model to any natural target system, which means that $N^{(n)}(x)$ in this model is not itself to be read as a population number. Instead, the process might best be interpreted as the iterative construction of a polynomial function of x. However, Devaney (1989, pp. 5–7) and the original Li and Yorke (1975) and Guckenheimer et al. (1977) motivate the construction of the iterated logistic model by reference to the "interesting" results obtained for the discrete logistic model (Section 2.3.2). The aim of this construction is therefore to create a novel mathematical object, whose properties can then be explored and which can be used in the exploration of related models. In contrast to the logistic models discussed in Section 2.3.1 and Section 2.3.2, the iterated model of the logistic equation is therefore a horizontal model (Section 2.2.2), which is constructed independently of empirical data or governing theory.

Behaviour of the Iterated Logistic Model
Devaney (1989) notes that the parameter r crucially influences the behaviour of the polynomial functions $N^{(n)}(x)$. In particular, for $r \leq 4$, each function will always map the unit interval onto itself, as illustrated in Fig. 2.2. In contrast, for $r > 4$, part of each function will always be mapped outside the initial domain, that is the unit interval. This is clearly visible in Fig. 2.3. This difference results in very different qualitative behaviours of the model in the two parameter regimes.

For $r < 3$, the behaviour is simple and mimics that of the continuous logistic model (Section 2.3.1). As shown in Fig. 2.2, successive self-iteration leads to an asymptotic approach of the model to a constant equilibrium value:

$$N_{eq} = \frac{r-1}{r}.$$

The equilibrium value is also the value of the single, non-trivial, iterative fixed point of the functions $N^{(n)}(x)$.

The parametric region $3 \leq r < 4$ is thus characterized by a transition from asymptotic behaviour to a dynamics characterized by oscillations of the functions $N^{(n)}(x)$ within the unit interval. Furthermore, the number of fixed points, that is those points at which the function cuts the line $N(x) = x$, will increase with each iteration (since the functions fill parts of the whole unit interval with an increasingly dense packing of peaks and troughs). The fixed points of $N^{(k)}(x)$ are called the periodic points of $N^{(n)}(x)$ with period k.

The least positive k for which $N^{(k)}(x) = x$ is called the prime period. The proliferation of periodic points in this regime can be linked to Sarkovskii's theorem, which predicts the number of prime periods present in a set of periodic points. Devaney (1989, pp. 60–67) discusses the implications of the fact that for $r = 3.839$, model (2.5) has a periodic point of prime period 3. According to Sarkovskii's theorem, this implies that the map also has periodic points of all other prime periods. In detail, a version of Sarkovskii's theorem reads (e.g. Devaney, 1989, pp. 60–62):

Theorem 1. Let $N: \mathbf{R} \rightarrow \mathbf{R}$ be continuous. Suppose [under self-application] N has a periodic point of prime period three. Then N has periodic points of all other prime periods [under self-application].

Therefore, for $r \geq 3.839$ the model has periodic points of all prime periods, that is during each new iteration new fixed points need to be created.

The behaviour of the iterated logistic model changes again for $r > 4$. For $r > 4$, the maximum value of $N^{(1)}(x)$ is $r/4$ and hence is greater than one. This implies that there exists a set of values A_1 which are mapped outside the unit interval under the first iteration. Looking at Fig. 2.3, we can see that A_1 is an open interval centred around $x = 1/2$. Devaney (1989, pp. 34–35) then shows that points that have been mapped outside the unit interval once, will be mapped monotonically to $-\infty$ through subsequent self-applications. This is well illustrated in Fig. 2.3. A different, more memorable, way of describing this behaviour is to say that points which have once "escaped" (p. 34) the unit interval, have escaped this interval for good.

It remains to consider the fate of the points staying in the unit interval. Following Devaney's notational convention, we will denote the set of points doing so up to an iteration n as:

$$\Lambda_n := I - \left(\cup_{k=0}^{n} A_k\right)$$

where A_k is the set of points escaping during the kth iteration. Each Λ_n is composed of the $2^{n+1} - 1$ closed intervals lying in between the open intervals of escaping points. The set of points remaining forever within the unit interval is therefore given by:

$$\Lambda = I - \left(\cup_{k=0}^{\infty} A_k\right). \tag{2.6}$$

It is intuitively graspable that the construction above could be described as "removing the middle". Devaney (1989, pp. 37–38) shows that it is indeed formally equivalent to the construction of a Cantor Middle Third set and that the resulting set Λ will be a Cantor set. This also implies that Λ will be fractal, that is self-similar under magnification.

Since all of the escaping points A_n are subsequently mapped to $-\infty$, while each iteration still maps some of the points in the intermediate intervals above the unit interval, the graph needs to have an intermediate maximum of subsequently escaping points on each of the closed intervals forming Λ_n and hence form a "hump" on each of those intervals. Accordingly, the nth iteration of eq. (2.5) crosses the line $N(x) = x$ at least 2^n times and so has at least a number $P_n = 2n$ of periodic points. Hence, one also finds:

$$P_n \rightarrow \infty, \; for \, n \rightarrow \infty. \tag{2.7}$$

That is, the model will display an infinite number of periodic points after an infinite number of iterations.

A further characteristic that Devaney (1989, pp. 36–37) proves for the iterated logistic model is also immediately apparent from Fig. 2.3: *the set of periodic points is dense* in Λ. The definition of a dense set is here the following standard one:

Definition 2. A subset U of S is dense in S if S is the closure of U, that is S consists of U and all its limit points.

Roughly speaking, under an infinite iteration of the "removing the middle" procedure described earlier, the disjoint sets forming Λ will become exceedingly small. However, since they are forced to each contain a periodic point, each such subset will eventually consist of this periodic point and an increasingly small "padding" of points on each side. For an infinite number of iterations, the set Λ will therefore "approach" the set of periodic points.

Devaney (1989, pp.47–50) uses a technically sophisticated proof (relying on symbolic dynamics and the topological conjugacy of the logistic map on Λ to a shift map) to show that $N^{(n)}(x)$ is *topologically transitive* on Λ. The definition of topological transitivity is here:

Definition 3. A map $N: J \to J$ is topologically transitive if for any pair of open sets $U, V \subset J$, there exists $n > 0$ such that $N^{(n)}(U) \cap V = \varnothing$.

This implies that, eventually, points from one arbitrarily small neighbourhood will be mapped into any other arbitrarily small neighbourhood.

In a similar procedure (again relying on the conjugacy to the shift map), Devaney (1989, p. 49) also shows that the iterated logistic model $N^{(n)}(x)$ on Λ possesses *sensitive dependence on initial conditions (SDIC)* for $r \geq 2 + \sqrt{5}$, in the sense that:

Definition 4. $N: J \to J$ has sensitive dependence on initial conditions if there exists $\delta > 0$ such that, for any $x \in J$ and any neighbourhood X of x, there exists $y \in X$ and $n \geq 0$ such that $|N^{(n)}(x) - N^{(n)}(y)| > \delta$.

Roughly speaking, this means that there is at least one point arbitrarily close to x which will eventually be separated from it by at least δ. Defining the distance δ requires the existence of a metric; but in our example it is the elementary one on the number line. Fig. 2.3 illustrates the existence of SDIC in the sense that that the continued narrowing of the intervals in Λ_n will lead to more and more points in the neighbourhood of a periodic point being eventually mapped throughout the whole unit interval. Hence any neighbourhood around a periodic point in Λ can be used to demonstrate SDIC.

Devaney (1989) initially only provides proofs of these last two properties for a growth rate $r \geq 2 + \sqrt{5}$ and only on the set Λ, which covers only a small fraction of the unit interval. Further proofs are provided for selected other cases: for example, Devaney (1989, pp. 50–51) shows that for $N(x) = 4x(1 - x)$, that is for the special case $r = 4$, where the maximum of the function is mapped exactly onto the unit line, the model has a dense set of periodic points, topological transitivity and SDIC on the whole unit interval; some other proofs for particular logistic functions are set as exercises for the reader.

This horizontal model derived from the logistic equation through iteration clearly has some very interesting mathematical properties. In

Section 3.2.3, I will show that its behaviour is paradigmatic for the definition of a kind of chaos, namely Devaney chaos, and that it is also chaotic according to a number of other chaos definitions (Section 3.4). Since the iterated model is a horizontal model, it is not directly tied to any target system and should therefore also not be expected to inform us about any real physical system. However, in Sections 4.2.2 and 4.3.1, we will see that the iterated logistic model is instrumental in the evaluation of the discrete logistic model: in particular, it is used to establish the fact that the aperiodicity displayed by the discrete logistic model can be viewed as having sufficient conditions in the model's nonlinear dynamics and discrete nature.

2.4   A LINEAGE OF LORENZ MODELS

In this section, I will describe the constructions and behaviours of two models based on the Lorenz equations. The first model is a discrete model constructed vertically by Lorenz (1963) as a means of describing convection in a heated layer of fluid (Section 2.4.1). This model is probably the best known model in chaos theory and the discovery of chaos in this model has become something of a founding legend for the field (e.g. for an account of the standard myth, Alligood et al., 1997, pp. 259–360). The second model is an iterated model (Section 2.4.1), constructed horizontally as a geometrical equivalent to the discrete Lorenz model. A third Lorenz model, which constitutes a hybrid between these two models, will be introduced in Section 4.5.

My description of the Lorenz models' constructions and behaviours will be less detailed than that of the logistic models in Section 2.3. This is due to the fact that a similar analysis of the Lorenz models would require the reproduction of much more physical and mathematical background material. Instead, my discussion will remain almost entirely conceptual.

2.4.1   Discrete Lorenz Model

The discrete Lorenz model aims to describe convective motion in the Earth's atmosphere. The general theory underlying the construction of the discrete Lorenz model, and hence the beginning of the lineage of Lorenz models, is that of hydrodynamics. However, Lorenz (1963) begins his model construction with an already simplified scenario: that of a two-dimensional layer of fluid with a constant temperature difference between the upper and the lower surface (Lorenz, 1963, p. 134).

Construction of the Discrete Lorenz Model

Lorenz (1963, pp. 133–135) constructs the model by successively simplifying the convection equations due to Saltzman, which provide the top-level theory for the model. In this case of model construction, the prepared description includes the assumption of free boundaries (p. 134); series truncation (pp. 134–135) and the neglect of all trigonometric terms (p. 135). The set of equations resulting from these modifications is that of the well-known Lorenz equations:

$$\frac{dX}{d\tau} = -\sigma X + \sigma Y,$$

$$\frac{dY}{d\tau} = -XZ + rX - Y,$$

$$\frac{dZ}{d\tau} = XY - bz, \tag{2.8}$$

where X is proportional to the intensity of convective motion in the layer; Y is proportional to the temperature difference between ascending and descending currents; Z is proportional to the deviation of the vertical temperature from linearity and the independent variable τ is a dimensionless time-measure. The equations contain three constants: r, σ and b.

Since this system cannot be solved analytically for explicit functions X (τ), $Y(\tau)$ and $Z(\tau)$, the construction of the model contains the further steps of discretization and numerical integration (Lorenz, 1963, pp. 133–134; p. 136). The discrete Lorenz model is therefore a numerical model, that is a model whose general set of equations as derived from the governing theory is continuous but which needs to be discretized to allow numerical integration of this set of equations. The final discrete model is not displayed on paper; however, Lorenz (1963, p. 136–137) summarizes its computational implementation in the following way:

> We have used the double-approximation procedure for numerical integration [...]. The value $\Delta\tau = 0.01$ has been chosen for the dimensionless time increment. The computations have been performed on a Royal McBee LGP-30 electronic computing machine.

This is next followed by Lorenz's (1963) tabulated X, Y, Z-values for the first 6,000 time-steps.

Behaviour of the Discrete Lorenz Model
Two plots of the behaviour of the discrete logistic model are shown in Fig. 2.4. In the following, the term "trajectory" will be used to describe the development of the vector function $(X(\tau)$, $Y(\tau)$, $Z(\tau))$. While the term "trajectory" indicates a continuous development, it should be kept in mind that these functions have only been evaluated at discrete values of τ. Similarly, plots of the model's output in Lorenz (1963) use continuous lines, while the output would actually be similar to the discrete points shown on plots of the discrete logistic model. My own Fig. 2.4 therefore shows such discrete plots but does not differ in other qualitative aspects from the behaviour described by Lorenz (1963).

Like the logistic models discussed in Section 2.3, the Lorenz model shows qualitatively different behaviours for different parameter values of r, σ and b. For the majority of combinations of these values, the model's trajectories regularly oscillate around two unstable equilibrium points. In the model system, this corresponds to the formation of steady convective rolls.

However, in the parameter regime $\sigma = 10$, $b = 8/3$ and $r > 0$, the model's behaviour changes and it begins to show aperiodic changes in the direction of its oscillations around the equilibrium points. Fig. 2.4 illustrates the model's output in this regime. Alligood et al. (1997, p. 365) describe the behaviour of the model in the following way:

> A trajectory will appear to spiral out around one equilibria [sic]...until its distance from this equilibrium exceeds some critical distance. Thereafter, it spirals around the other equilibrium with increasing amplitude oscillation until the critical distance is exceeded again.

This aperiodic cycling patterns will also be quantitatively different for different initial values of X, Y and Z. In the model system, this aperiodic regime corresponds to the convective flow in the layer reversing its direction at aperiodic intervals.

Despite the aperiodic cycling around the equilibrium points, the trajectories appear to be confined to a finite butterfly-shaped region of the XYZ-phase space (Fig. 2.4a). This region is called the Lorenz attractor. An attractor is a region in phase space to which all points or

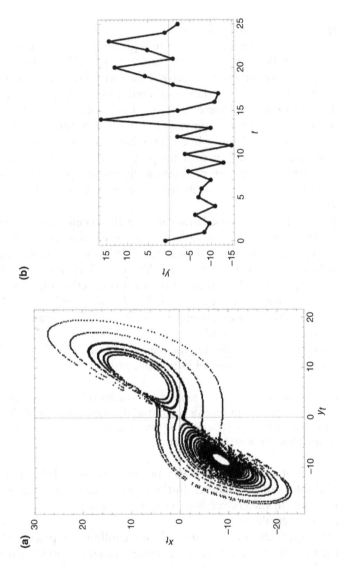

Fig. 2.4 Behaviour of the discrete Lorenz model. (a) Aperiodic oscillations around the two equilibrium points in the XY-plane. (b) Aperiodic development of the variable $X(\tau)$

trajectories of a function will eventually converge. We have already encountered an attractor in the discussion of the iterated logistic model (Section 2.3.3): The Cantor set Λ is an attractor for the model's set of fixed points.

Further detailed analysis of the model (e.g. Lorenz, 1963, pp. 138–140; Alligood et al., 1997, pp. 366–370) shows that the behaviour of the discrete Lorenz model also undergoes phases of infinite periodicity and extinction similar to the one found in the discrete logistic model (Section 2.3.2). In Section 3.4, I will show that the discrete Lorenz model has been found to be chaotic according to several definitions of chaos.

While the butterfly-shaped attractor appears robust in computer simulations of the discrete model, formal evidence for its existence was only provided through extensive investigative work with two horizontally constructed models (Section 2.4.2 and 4.5). The models were also used to provide evidence for the fact that the discrete nature of the model, which in this case is fictional, is not a necessary condition for the display of chaotic behaviour, that is, there exist other sufficient conditions for this behaviour. The most prominent attempts to gather this evidence through the work with related horizontal models will be discussed in detail in Section 4.5. Since the Lorenz models are considerably more complex than the logistic models, the investigation of the original model through work with others in the lineage has proven correspondingly more difficult. In Sections 4.3.1 and 4.3.3, I will argue that both the need to rule out discreteness as a necessary condition for chaos in the model as well as the technically complexity of the modelling work required to do so are representative for the general evaluation of numerical models in chaos theory.

2.4.2   Iterated Lorenz Model

The construction of the iterated Lorenz model by Guckenheimer (1976) can be viewed as an attempt to establish mathematical constraints on the conditions under which functions will display trajectories that mimic the figure-eight behaviour displayed by trajectories in the discrete Lorenz model (Section 2.4.1, Fig. 2.4). Under this interpretation, the model's construction is therefore a further example of the horizontal construction of models in chaos theory.

In order to avoid reviewing a great amount of technical material on differential topology, my account of the iterated Lorenz model's construction will only sketch out its most important conceptual steps. A discussion of the geometric attractor aimed at a non-expert, albeit still mathematically literate, reader can be found in Viana (2000). A very good illustration of the topological set-up of the iterated Lorenz model can also be found in Viana (2000, p. 12).

Construction of the Iterated Lorenz Model
Guckenheimer's (1976, pp. 370–372) construction of the model consists of specifying a set conditions for the behaviour of the trajectories of a hypothetical function $f\colon \mathfrak{R}^2 \to \mathfrak{R}^3$ in a three dimensional phase space, which he then shows to be sufficient conditions for both the existence of a butterfly-shaped attractor as well as for a number of other interesting properties. The construction can be divided into three conceptual steps:

- **Placing of equilibrium point:** Firstly, an equilibrium point O is (arbitrarily) defined. Then, relative to this point O, two planar regions are defined: A square Σ; and a plane Γ, which is perpendicular to Σ, divides Σ into two halves; and contains O.
- **Prescription of first return maps on Σ and Γ:** Secondly, some constraints are put on how trajectories that have cut Σ and Γ once will cut these planes on their next return. That is: conditions are put the first return maps for these trajectories on Σ and Γ. To do this, two convex triangles are inscribed onto Σ and it is required that trajectories that cut Σ within one of these triangles once will cut the square within one of the triangles again the next time they return. Additionally, these trajectories are required to change directions with respect to Σ before returning, which establishes the desired figure-eight pattern. A hypothetical trajectory running through O is required to next cut the square Σ at the tip of one of the triangles, while also looping round. Since there are two possible such trajectories – one to each tip, in the following denoted as γ_+ and γ_- – these "boundary" trajectories and O form an unstable manifold, which is locally orthogonal to Γ.
- **Prescription of stable and unstable manifolds:** The tendency of the trajectories to stay on and converge onto the attractor is then enforced by prescribing additional stability conditions. The figure-eight pattern of the trajectories has already been established by defining the unstable manifold through O by means of the two trajectories γ_+ and γ_-. It is now additionally specified that Γ is a stable

manifold of O, that is, all trajectories on Γ will eventually converge to O. This ensures that all trajectories in the vicinity of O will eventually be forced onto the attractor.

During the construction of the iterated Lorenz model, no single function f is explicitly postulated. Rather, by specifying the tendencies of trajectories in different regions of phase space, Guckenheimer (1976, pp. 370–372) provides minimal instructions for the iterative generation of trajectories through the iterative determination of those areas on Σ that the trajectory will be forced to pass through on its next return to the plane. As a matter of fact, after establishing the general area of the attractor, no trajectories are ever iterated beyond their first return to Σ.

The construction of the iterative Lorenz model is an interesting illustration of the use of horizontal models in chaos theory. The model is clearly not constructed from top-level theory and bottom-level empirical data (Section 2.2.1) but consists of an assembly of conditions on the phase space structure of a model sufficient to force any model obeying the conditions to mimic certain aspects of the behaviour of another, given, model. The construction of the model itself is part of its investigatory use since it serves to establish (i) that it is possible to analytically construct an attractor of the same geometric form as the discrete Lorenz model might display and (ii) which conditions placed on the model's dynamics are sufficient for this. Importantly, the iterated Lorenz model is not a numerical model but consists of iteratively constructed continuous trajectories. Accordingly, if the results from the iterated Lorenz model transfer to the discrete Lorenz model, then the discreteness of the latter model, which is an artefact of the numerical integration, does not have to be seen as part of the sufficient conditions for the appearance of chaos (of some definition), that is a conditional $C \rightarrow B$ that does not contain discreteness in the antecedent can be determined. The use of the iterated logistic model in the evaluation of the discrete logistic model will be discussed in Section 4.5. Since the models are both relatively complex, this investigation involves the construction of a third Lorenz model, the rigorous Lorenz model, whose construction as a hybrid model from the iterated and discrete model will also be outlined in Section 4.5.

Behaviour of the Iterated Lorenz Model
The aim of the construction of the iterative Lorenz model is to show that a model thus constructed would feature a figure-eight shaped attractor, that

is that a region neighbouring the figure-eight shaped region specified by O, Σ and Γ could be found from which all trajectories would converge onto the attractor. The existence of this neighbourhood is formally proven by Guckenheimer (1976, p. 372–376).

Guckenheimer (1976, pp. 376–382) establishes two further important features of the model: that (i) the model has a form of topological transitivity similar to that of Definition 3 (Section 2.3.3) in that there exists at least one trajectory which visits all possible subregions of phase space; and that (ii) there exists subregions of the triangles on Σ, which are invariant under the first return map of the given function f. Invariant here implies that any two points that are located in a specific, leaf-like part of the invariant region will both be mapped to another leaf-like subpart of the invariant region by the first return map. Guckenheimer (1976, pp. 379) shows that, with the ongoing iteration of such trajectories, points in the invariant regions of the first return maps will move closer together and eventually have a Cantor-like fractal structure similar to the fixed point set of the iterative logistic model (Section 2.3.3). It can then be shown (e.g. Viana, 2000, p. 10) that, on these invariant regions, the attractor exhibits SDIC according to Definition 4: the distance between two trajectories, which are initially close together, but end up cutting Σ in different invariant regions will continuously increase with each iteration, since the regions are becoming successively narrower with each iteration.

Despite the fact that the iterated Lorenz model was constructed to mimic the shape of the discrete Lorenz model (Section 2.4.1), its properties are also very similar to that of the iterative logistic model. In fact, Hirsch et al. (2004, pp. 314–318) show that the iterated Lorenz model possesses all of the relevant properties of the iterated logistic model (Section 2.3.3) and that it is therefore also Devaney chaotic (Section 3.4.1). In Section 3.4, I will show that it has also been diagnosed as exhibiting several other types of chaos.

2.5   CONCLUSION

In the first part of this chapter, I reviewed the construction and evaluation of vertical and horizontal models (Section 2.2). Vertical models (Section 2.2.1) are constructed from top-level theory and bottom-level empirical target systems, through the provision of prepared descriptions. Their evaluation therefore consists of a comparison of the model systems (i.e. the hypothetical concrete system described by the model) with the target system. Following Suarez (2013), I have conceptualized the evaluation of vertical models as the evaluation of the transference of conditionals C → B between the model and target system. In this context, I have also introduced a notion of model faithfulness (Definition 1), which states that a behaviour B is modelled faithfully if the antecedent C in an appropriate conditional C → B does not contain any fictional parts, that is if the behaviour B is conditionally isolated from any fictional effects arising from the model's construction. A further advantage of this analytic framework, which I hope will become apparent in the following chapters, is its natural separation between the conditions C, which are sufficient for the appearance of a behaviour in the model B, and the criteria for the diagnosis of this model B.

Following Bokulich (2003), I maintained that – besides vertical models – there is a second important class of models in chaos theory: horizontal models (Section 2.2.2). Horizontal models are not constructed from theories and empirical data but are mathematical variations of existing models. Such horizontal spin-offs can lead to the formation of lineages of differently constructed but related models. The evaluation of horizontal models does not entail an assessment of their informativeness towards a natural target system, but consists simply in an exploration of the mathematical properties of models in their lineage. In Section 4.4, I will show that, for horizontal models in chaos theory, this investigative function can be specified even more precisely: we will see that they are often instrumental in establishing the conditionals C → B to be transferred during the evaluation of a related vertical model.

In the second part of the chapter, I illustrated the horizontal and vertical construction of models in chaos theory through two case studies. In Section 2.3, the construction and behaviour of a lineage of three models based on the logistic equation was discussed. Two of these models, the continuous logistic model (Section 2.3.1) and the discrete logistic model 2.3.2, are developed as vertical models to describe the dynamics of overlapping and non-overlapping populations, respectively (May, 1974). A third, horizontally constructed, logistic model (Section 2.3.3) is presented by, for example, Li and Yorke (1975), Guckenheimer et al. (1977) and Devaney (1989): viz. the iterated logistic model, which is generated through successive self-application of the logistic equation.

In Section 2.4, I showed that there exists a similar lineage of models based on the Lorenz equations. The discrete Lorenz model (Section 2.4.1) is constructed vertically as a representation of atmospheric convection

(Lorenz, 1963). Based on this vertical model, Guckenheimer (1976) constructs a horizontal model (Section 2.4.2), consisting of the iterative generation of trajectories of a function f, which are deliberately constrained to move on an attractor similar to that observed in the discrete Lorenz model.

In Chapter 3, I will develop a framework for the analysis and comparison of different chaos definitions. It will be shown that each of the models presented in this chapter is chaotic according to several different chaos definitions. The models introduced in this chapter will also serve as case studies in Chapter 4, where I will address the evaluation of chaotic models.

Chaos Criteria and Definitions

Abstract I will maintain that chaos definitions are twofold decomposable into combinations of five core criteria (determinism, transitivity, periodicity, aperiodicity and SDIC) and into different embodiments of these criteria. The different embodiments of each criterion and the relationships between different criteria will be discussed. I will then show that five prevalent definitions of chaos can be analysed in this compositional framework.

Keywords chaos · definitions of chaos · definitions in science

3.1 INTRODUCTION

In this chapter, I will analyse the criteria for, and definitions, of chaos. The chapter has two main theses: (i) I will maintain that there are five main criteria for diagnosing chaos – one dynamical criterion and four phenomenological criteria; and (ii) that studies on chaos use various combinations and embodiments of these criteria to define chaos. I therefore view definitions of chaos as twofold "decomposable": into different combinations of criteria and into specific embodiments of these criteria.

This view implies that there naturally exists a large number of different chaos definitions, which – while grounded in the same four criteria – can differ in their specific requirements. Such differences can include both the requirements of different combinations of criteria as well as differences in the technical specifications of a given criterion, that is with respect to the

© The Author(s) 2017
L.C. Zuchowski, *A Philosophical Analysis of Chaos Theory*, New Directions in the Philosophy of Science, DOI 10.1007/978-3-319-54663-6_3

domain of applicability or the type of functions it is targeted towards. On the one hand, a model (or natural system) can therefore often be diagnosed as having several different kinds of chaos, that is the extensions of many of the chaos definitions discussed here overlap to a considerable degree. On the other hand, the variety of different technical specifications of the requirement for chaos allows for a large number of different models to be given the general label "chaotic". As I will illustrate in the case study of the diagnosis of chaos in the lineage of logistic models (section 3.2) this means that often almost all models in a lineage (Section 2.2.2) can be called chaotic, albeit in technically somewhat different ways. One of the epistemic role of the coexistence of these different definitions of chaos therefore seems to be to allow for a degree of cohesion in the characterization of the different kinds of models used in the field while recognizing the difference in the technical realization of these models.

The chapter is to be understood as an exercise in rational reconstruction: it aims to provide a conceptual framework in which different chaos definitions can be analysed and compared to each other. My motivation for using this particular framework is purely practical: it appears to me that it allows us to display the requirements made by different chaos definitions clearly and (to a degree) comparatively. In line with the general aims of rational reconstruction, I thereby aim neither to be descriptively correct, that is to claim that I have discovered how scientists in practice construct definitions of chaos, nor to be normatively prescriptive, that is to claim that scientists should construct chaos definitions in this particular way. Instead, I view the decompositional analysis presented here as a useful tool of conceptual analysis, which will allow the philosopher of science to gain greater insight into both the use of as well as the formal structure of and relationships between different definitions of chaos. I also do not claim that this framework is the only possible one for the rational reconstruction of chaos definitions. However, I am convinced that it is a serviceable one and hope to demonstrate its virtues in this chapter.

The existence of many different definitions of chaos is generally recognized by both philosophers and practitioners. However, the philosophical literature on chaos usually either focuses on describing the advantages or disadvantages of one or two such definitions (e.g. Batterman, 1993; Werndl, 2009c) or assumes that there exist no fundamental differences in the first place (e.g. Berger, 2001, p. 40). To my knowledge, there have been no attempts at a detailed survey of the chaos definitions in use. Smith (1998, Chapter 10) comes closest to such a project but does not go beyond recognizing the existence of many different chaos definitions;

ultimately the author concludes that defining chaos should be a task for practitioners and not for philosophers (p. 184). While I agree that it is not philosophically fruitful to search for the one "correct" definition of chaos, I think that comparatively analysing the structure and use of different chaos definitions is an appropriate task for the philosopher of science. In contrast to Smith (1998), who places the discussion of how chaos is defined at the end of his book, I view this analysis as preliminary work to the evaluation of chaotic models (Chapter 4). Therefore, my work here aims to complement the existing work by attempting such a detailed longitudinal study (albeit not aiming at a complete survey).

3.1.1   Sufficient Conditions and Criteria for Chaos

Viewed in the general analytic framework of the inferential account of model evaluation by Suarez (2013), which I have introduced in Section 2.2.1, this chapter is solely concerned with the analysis of description of the behaviour B in the conditionals C → B to be determined and transferred from a model to the target system. My account therefore strongly distinguishes between criteria for the diagnosis of this behaviour B, which are collected in the definitions of chaos analysed in this chapter, and the conditions C, which are seen as sufficient conditions for this behaviour to occur in the model.

The sufficient conditions C for the occurrence of behaviour B in a model or target system and the criteria used to define this behaviour are usually formally distinguished by the latter having to be both necessary and sufficient for the behaviour to be diagnosed as such, that is definitions are viewed as tautological in a logical sense. This formal difference is usually not articulated explicitly in existing studies and (it seems to me) this lack of articulation can occasionally lead to confusions in discussions on both how chaos should be defined and what should be seen as sufficient conditions, that is as explanatory in a mild interpretation of the word, for it in a model (e.g. as demonstrated in the very interesting transcripts of a number of podium discussions on the definition and implication of chaos collected in Schurz, 1996).

Since the difference between sufficient conditions for the occurrence and criteria for the definition of chaos is usually not articulated explicitly, my division of concepts into these categories was guided by the use of particular notions in the existing discourse and the desire to achieve a division that allows me to dissect prevalent issues with the greatest possible clarity: that is by the twin demands of rational reconstruction to be true to the actual use of a concept in science while displaying it in a way that "slices nature at its joints". I do not think that my division here has to be normative. However, I hope that the clear exhibition of the sufficient conditions for the occurrence of chaos and the criteria used for its definition will serve as a good starting point for a more nuanced discussion of alternative approaches to the definition of chaos: for example, the one by Smith (1998, chapter 10), who proposes to consider the existence of a kneading sequence, which I think is usually seen as a sufficient condition for the occurrence of chaos, and have classified as such (Section 4.3.1), as a defining criterion.

One of the main results of my analysis of the use of chaos definitions is that, while this generally holds true for the label "chaotic" as well (section 3.3 and 4.3.1), chaos definitions usually require an additional dynamical (pre-)criterion: namely, determinism (section 3.3.1). I will argue in section 3.3.1 that the relevant discourse indicates that the existence of a deterministic set of equations is used as a criterion for the definition of chaos rather than a sufficient condition for the occurrence of chaos. The use of this dynamical criterion then leads to one of the most striking epistemic features of chaos theory as a scientific field: it leads to the need to diagnose a dynamical property from the phenomenology of a model or a system. I will argue that the difficulty in accomplishing this is one of the main features of the use and evaluation of model in chaos theory and one of the main reasons why – in contrast to the astronomy case study discussed by Suarez (2013) – the evaluation process in chaos theory is heavily skewed towards step (ii) in the evaluation process, that is towards the determination of the existence of chaos in nature. The fact that this epistemic feature of chaos theory can be displayed clearly in my analysis seems to be one of the advantages of the framework used here.

Similarly, I will be able to give a clear exposition of another thorny issue in philosophical discussion of the definition of chaos: namely, whether a lack predictability should be seen as criterion for the definition of chaos. I will thereby be able to provide further evidence towards the view that this should not be the case (e.g. Batterman, 1993; Werndl, 2009c).

3.1.2   Outline of the Chapter's Content

I will begin this chapter (section 3.2) by returning to one of the case studies introduced in Chapter 2 and will discuss the diagnoses of chaos in the three logistic models (Section 2.3). Thereby, I will show that the continuous logistic model (Section 2.3.1) has not been found to be chaotic (section 3.2.1). The discrete logistic model (Section 2.3.2) is found to match three possible criteria for the diagnosis of chaos, namely that of determinism, aperiodicity and SDIC (section 3.2.2). Accordingly, it has been found to be chaotic by May and Osler (1976), using a chaos definition that I will call "stochastic chaos" (section 3.4.4). In section 3.3 and section 3.4, it will also become apparent that the model also fulfils embodiments of the criterion of transitivity and is chaotic according to a large number of other chaos definitions.

The iterated logistic model (Section 2.3.3) matches four possible criteria for the diagnosis of chaos: it possesses embodiments of transitivity; of SDIC and of periodicity (section 3.2.3). In section 3.2.3, I will discuss how Devaney (1989) diagnoses the model as being chaotic, using his own definition of chaos based on these three criteria. In section 3.4, it will become apparent that the model is also chaotic according to several other chaos definitions.

Thus section 3.2 provides descriptions of two examples of how chaos definitions are constructed and used in practice. In addition, the section also introduces the five criteria that I consider to be constitutive of virtually all existing chaos definitions: determinism (section 3.3.1), which is distinguished from the other four criteria as presenting a dynamical pre-criterion, transitivity (section 3.3.2), periodicity (section 3.3.3), aperiodicity (section 3.3.4) and SDIC (section 3.3.5). The criterion of determinism is a dynamical criterion, that is it puts demands on a chaotic model's equations, while the four other criteria are phenomenological criteria, that is they require a chaotic model's behaviour to display certain features.

In section 3.3, I will discuss these five criteria in detail. It will become apparent that the criteria are "similarity categories" and can assume many different formal or semi-formal "embodiments". An example of this has already been encountered in Sections 2.3.3 and 2.4.2, where we have found the notion of transitivity to be defined differently for the iterated logistic model and the iterated Lorenz model, respectively. Similarly, the criterion of SDIC can be further specified through Definition 4 or by requiring positive Lyapunov exponents (section 3.3.5). The most common embodiments of each criterion will be discussed. I will also outline the relationships between the criteria. It will thereby become apparent that SDIC has a specific relationship with the three other phenomenological criteria: on the one hand, depending on the embodiments used for these criteria, SDIC can be viewed as an implication of combinations of other criteria; on the other hand, attempts to define chaos through requiring a lack of practical predictability usually view SDIC as the sole constituting criterion of chaos. In section 3.3.5 in this chapter and later on in Section 4.3, I will briefly argue that the definition of chaos as this particular kind of unpredictability is problematic: both because it leads to an internal conflict with the criterion of determinism as well as because it seems difficult to maintain that chaos can ever be modelled faithfully if such a definition is used.

In section 3.4, I will demonstrate how the most prevalent chaos definitions are composed of different combinations of the five core criteria. Five chaos definitions will be analysed: Devaney chaos (section 3.4.1); (definition of chaos as) mixing (section 3.4.2); (definition of chaos through) positive Lyapunov exponents (section 3.4.3); (definition of chaos through) strange attractors; and stochastic chaos (section 3.4.4). It will be shown that analysing the definitions as consisting of different combinations of the four criteria makes it possible to at least heuristically compare all four definitions of chaos. The section will demonstrate that the extensions of all of these definitions overlap considerably. If this is not the case, that is, if there exist models that are diagnosed as being chaotic according to one definition but not according to another, then this can often be traced back to the need to make a definition applicable to a particular class of models. Accordingly, the coexistence of many different definitions of chaos does not appear to highlight any large conceptual divides in how the core concept is construed but rather seems to be a consequence of the complex interplay of different models in chaos theory.

3.2   Diagnosis of Chaos in the Logistic Lineage

In this section, I will describe in detail how two of the logistic models introduced in Section 2.3 have been found to be chaotic. As a matter of fact, it will become apparent that there are two definitions of chaos, stochastic chaos and Devaney chaos, which have been developed specifically to describe the behaviour of the discrete and iterated logistic model, respectively (section 3.2.2 and section 3.2.3). The models can therefore be seen as exemplars of these definitions. The case studies in this chapter will focus on these stereotypical diagnoses of chaos, which are also historically the most relevant ones. In section 3.4, we will see that the discrete and iterated logistic model are also chaotic according to most other chaos definitions.

The two chaos definitions exemplified by the discrete and iterated logistic model can be viewed as being based on different combinations of five possible criteria for chaos: determinism, transitivity, SDIC, periodicity and aperiodicity. In section 3.3, different formal embodiments of these criteria and relationships with each other will be discussed. I will then show that combinations of these criteria are also required by other prevalent chaos definitions (section 3.4).

3.2.1   No Chaos in the Continuous Logistic Model

The continuous logistic model (Section 2.3.1) displays monotonic behaviour and has therefore not been labelled chaotic. This demonstrates that models derived from the same general theory can either be chaotic or not be chaotic, highlighting the fact that chaos appears to be a property of models rather than of theories.

3.2.2   Stochastic Chaos in the Discrete Logistic Model

This case study focuses on the diagnosis of chaos in the discrete logistic model: a diagnosis made by May and Osler (1976). The authors justify their description of the model's behaviour as deserving the (then novel) label of "chaotic" by stressing that it fulfils three criteria: the model's equations are deterministic; its behaviour is aperiodic and it possesses a form of SDIC. Their treatment of each of these criteria will be described below.

Determinism
May and Osler (1976) stress that the models they are analysing in their study all fulfil one dynamical criterion: they are deterministic, that is, the model's equations are such that each initial value leads (in principle) to a unique solution. The authors (p. 573) also make it clear that determinism is a required rather than incidental feature of the kind of chaotic behaviour they wish to describe, which crucial differentiates "chaos" from "noise":

> ...[T]here is still a tendency on part of most ecologists to interpret apparently erratic data as either stochastic 'noise' or random experimental error.

There is, however, a third alternative, namely, that wide classes of deterministic models can give rise to apparently chaotic dynamical behaviour. It is this third possibility which we elaborate in this paper.

The first criterion for chaos required by May and Osler (1976) is therefore determinism. This criterion is clearly fulfilled by the discrete logistic model (2.4). In section 3.3.1, I will describe the specific embodiments of this criterion, including the somewhat misleading stipulation that the dynamics of chaotic models needs to be "simple", in detail.

Aperiodicity
May and Osler (1976, e.g., p. 582) use the term "chaotic" synonymously with "aperiodic", thereby following the convention introduced by Li and Yorke (1975). They describe the behaviour of the discrete logistic model for $r > 2.57$ as "chaotic" and this parameter range as the "chaotic regime".

This general criterion of aperiodicity is then further specified as requiring indistinguishability from randomness, that is as requiring apparent stochasticity. The underlying conception of randomness here is that of a Bernoulli process: roughly, a process that consists of a sequence of trials with a given set of outcomes, whereby the probability for obtaining a given outcome is the same during each trial. The dynamics of a Bernoulli process is therefore not deterministic. The most prominent example of a Bernoulli process is a sequence of coin tosses, where the probability of obtaining the outcome of "heads" or "tails" is $1/2$ during each toss. May and Osler (1976, p. 585) then require that the phenomenology of a chaotic model is undistinguishable from the phenomenology of a Bernoulli process, for example, a recording of the outcomes of a sequence of coin tosses:

> If we agree to count any point that lands in the left [lower] half of the interval as 0 and any point in the right half [upper] as 1 (i.e., round to the nearest integer), then either graph will generate a sequence of 0's and 1's that are indistinguishable from a Bernoulli process consisting of a sequence of coin tosses.

Formally defining the notion of a Bernoulli process is technically difficult and requires the formulation of an appropriate probability measure on the space of possible outcomes of a system (e.g. Werndl, 2009a, p. 233). In section 3.3.1, I will briefly discuss the notion of a Bernoulli process as a paradigmatic indeterministic system in more detail; in section 3.3.4 I will briefly describe which phenomenological features are characteristic for a Bernoulli process and are therefore often used as embodiments of the criterion of aperiodicity for chaotic systems as well. However, for the majority of the discussion in this book, it is sufficient to understand the notion of apparent stochasticity on a purely conceptual level as phenomenological indistinguishability from a process that is generally accepted as "truly" stochastic. Most often, the stochastic comparison process is that of a coin toss: that is apparent stochasticity requires that a representation of a model's outcome can be found in which this outcome is indistinguishable (relative to some criteria) from a suitable representation of the outcome of a sequence of coin tosses.

Sensitivity to Initial Conditions
May and Osler (1976, pp. 586) note that the apparent stochasticity of the discrete logistic model also entails a second property: since stochastic sequences are intuitively maximally different from each other and each initial condition creates a different stochastic sequence, the model possesses a form of SDIC. In contrast to the formal embodiment of SDIC in Definition 3 (Section 2.3.3), May and Osler (1976, pp. 586) discuss SDIC in general terms, noting that such tiny fluctuations in the initial values will "fuzz out" the potential number of values that could correspond to later iterations of the model. However, it has been formally established that the phenomenology of a Bernoulli process also has the characteristic of SDIC; the assumption that this property will be implied by the apparent stochasticity of the discrete logistic model is therefore consistent with the definition of apparent stochasticity as phenomenological indistinguishability from a Bernoulli process.

Since it is based on indistinguishability from stochasticity, I have named the chaos definition used by May and Osler (1976) "stochastic chaos". The definition requires forms of the general criteria of determinism and aperiodicity. In the embodiment used by May and Osler (1976), the criterion of aperiodicity implies the satisfaction of a third criterion for chaos, namely SDIC. In section 3.3.4, we will see that this is the case for many, but not all, embodiments of this criterion. The discrete logistic model is also chaotic according to a number of other chaos definitions: in particular, according to definitions of chaos as mixing (section 3.4.2) and definitions of chaos in terms of the existence of positive Lyapunov coefficients (section 3.4.3).

3.2.3   Devaney Chaos in the Iterated Logistic Model

In this case study, I will outline Devaney's (1989) diagnosis of the iterated logistic model as chaotic. An updated, but not fundamentally different, discussion of this material can be found in Hirsch et al. (2004, chapter 15).

Determinism
While Devaney (1989) does not stress determinism as a criterion for chaos as strongly as May and Osler (1976) do, he emphasizes (e.g. p. vii; p. 2) that his investigation is only concerned with properties of deterministic maps, that is that it excludes the investigation of any probabilistic processes. Accordingly, I think that one is justified to assume that determinism (section 3.3.1) is an unarticulated dynamical requirement for Devaney's (1989) definition of chaos. This implies that, even if the three phenomenological criteria discussed below were to be fulfilled by a probabilistic system, this system would not be considered to be chaotic. While this implication has not been expressed by Devaney (1989) or Hirsch et al. (2004) directly, it appears to be true to the spirit of their writing.

SDIC, Transitivity and Periodicity
Devaney (1989, p. 50) gives a formal definition of chaos, which specifies three criteria to be fulfilled conjointly by a chaotic map N:
   Definition 5. Let V be a set. $N: V \rightarrow V$ is said to be chaotic if

1. N has sensitive dependence on initial conditions; and
2. N is topologically transitive; and
3. periodic points are dense in V.

The formal definitions of each of the three criteria required by Devaney (1989) have been discussed in detail in Section 2.3.3. These definitions constitute embodiments of the general chaos criteria of SDIC (section 3.3.5), transitivity (section 3.3.2) and periodicity (section 3.3.3). It is notable (e.g. Hirsch et al., 2004, p. 338) that the definition combines two topological embodiments with a metric one: the requirements on the set of periodic points and the transitivity of the functions are topological in nature, while the embodiment of SDIC, as given by Definition 4, refers to pairs of points on the polynomial functions generated by the self-application process and requires the notion of a metric the space V, that is a reference system for the definition of a way to measure the distance between points.

However, Banks et al. (1992) formally show that, in Definition 5, the criterion of transitivity together with the criterion of dense periodicity implies the existence of a metric on V such N has SDIC with respect to this metric. Accordingly, Definition 5 could be reformulated as a purely topological definition of chaos. The definition is also another example of a chaos definition which has SDIC as implication rather than independent criterion (section 3.3.5).

We have already seen in Section 2.3.3 that the iterated logistic model fulfils all of the criteria required in Definition 5 and is therefore Devaney chaotic. In fact, the discussion in Devaney (1989, pp. 31–53) leaves no doubt that Definition 5 has been developed specifically to describe the properties of the iterated logistic model. The domain of applicability and the specific embodiments of the criteria for chaos in this definition are therefore also geared towards models which are similar to the iterated logistic model, that is iteratively constructed continuous functions. In contrast, the definition of stochastic chaos used by May and Osler (1976) specifically fitted the discrete nature of the discrete logistic model.

It is possible to diagnose the iterated logistic model as apparently stochastic in comparison to a Bernoulli process in the specific case of $\mu = 4$ (Section 2.3.3) or with a probability measure specifically defined to fit the disjoint domain of the set of fixed points Λ. However, such diagnosis require specific reformulations of the notion of apparent stochasticity and the use of a particular probability measure on the space of possible outcomes.

The two definitions discussed here, stochastic chaos and Devaney chaos, are therefore not directly applicable to both models but are specific to one of the models in the lineage. However, they both capture the properties that will be important in the further evaluation of the models in this lineage. In particular, Devaney chaos summarized those properties of the iterated logistic model, which will later on be crucial to the investigative use of this horizontal model during the evaluation of chaos in the discrete logistic model (Section 4.2.2). While the two definitions therefore require different combinations of criteria for chaos in different model specific embodiments, these difference do not seem to indicate a fundamental division in the conceptualization of chaos but rather a complementary fine-tuning of the concept to best fit a given model. In section 3.4, I will also show that there are many definitions of chaos which are directly fulfilled by both models, including definitions of chaos as mixing (section 3.4.2) and in terms of the existence of positive Lyapunov coefficients (section 3.4.3).

3.3   Dynamical and Phenomenological Criteria for Chaos

During the analysis of the initial diagnosis of chaos in the lineage of logistic models (section 3.2), five general diagnostic criteria for chaos were introduced: determinism, transitivity, periodicity, aperiodicity and SDIC. As discussed in section 3.2, the criterion of determinism is unusual in that it is a dynamical (pre-)criterion. In section 3.3.1, I will discuss the fact that this dynamical criterion has a unique epistemic function: it differentiates some conceptualizations of chaos from that of true stochasticity. In Section 4.3.2, I will argue that the use of this criterion also possess unique challenges for the evaluation of chaotic models. In the case study of the logistic lineage, the two chaos definitions used to diagnose chaos each (i) require different combinations, and (ii) use different embodiments, of these criteria.

The second point (ii) is most apparent in the requirement of SDIC in stochastic and Devaney chaos, respectively: while Devaney (1989) provides a formal definition of the criterion (Definition 3, Section 2.3.3), May and Osler (1976) consider SDIC an implication of the apparent stochasticity of their model. They provide a procedural definition for the notion of apparent stochasticity, based on a comparison to a discrete Bernoulli process, and explain how apparent stochasticity leads to a "fuzzing out" of initially close values (section 3.2.2). The two embodiments of SDIC are therefore specifically geared towards the models they should be applied to and therefore differ in their domain of applicability, that is in targeting continuous and discrete models, respectively. Nevertheless, it is apparent that the two notions are similar enough for the reader to immediately recognize that they are embodiments of the same core concept.

I maintain that there are five such core criteria (as named above: determinism, transitivity, periodicity, aperiodicity and SDIC) and that virtually all chaos definitions can be given in terms of combinations of (embodiments of) these criteria. For the four phenomenological criteria, it will become apparent that large numbers of formalizations of each criterion exist. While being embodiments of the same general criterion, these formalizations can differ in their targeting of discrete or continuous models; their use of statistical or deterministic means; or other similar technical specifications. Therefore, the criteria appear to be best viewed as similarity categories, which each contain a number of different formal embodiments of the given concept. The five criteria will be described in detail in section 3.3.1– 3.4.4. For each criterion, I will discuss some of the most commonly used embodiments, without making any claim of having covered all existing forms of that criterion. The most prevalent embodiments for each criterion have also been collected in Table 3.1.

Viewing the five criteria for chaos as similarity categories allows us to reconstruct chaos definitions through a doubly decompositional process: first, we identify the combination of criteria required by the definition,

Table 3.1   The five criteria for chaos and some of their major embodiments

Determinism (dynamical)	– Deterministic equations. – "Simple" equations –…
Transitivity	– Topological transitivity – Ergodicity – Dense trajectories –…
Aperiodicity	– Apparent stochasticity – Local unpredictability – Uniform Fourier profile –…
Periodicity	– Dense set of periodic points – Fractal features – Attractors – Some element of regularity –…
SDIC	– Topological SDIC – Positive Lyapunov exponents – Global unpredictability –…

then we identify the specific embodiments given to these criteria. This variable structure of chaos definitions can be interpreted as increasing the epistemic functionality of chaos definitions: casting a criterion into a specific embodiment allows this requirement to be articulated for technically different models, for example, the discrete and iterated model in the logistic lineage. Similarly, using different combinations of criteria in different definition of chaos can in some cases express differences in the conception of the essence of chaos, but it is also a means to highlight properties that will be important in the investigations the model is used for. In the latter case, the use of multiple definitions of chaos based on different combinations of the five core criteria for different models in the same lineage can also be viewed as a means of adjusting the relevant definitions of chaos to best fit the specific epistemic role of a model. This aspect of the use of definitions in chaos theory will be illustrated in the case studies of the evaluation of models in the logistic and Lorenz lineages (Sections 4.2 and 4.5).

As outlined in section 3.1, I do not claim that this two-step process is an historically or psychologically accurate description of how chaos definitions are constructed in practice. Instead, I am proposing this doubly decompositional framework as a useful tool for the analysis of the structure and functionality of chaos definitions, that is as a means of rational reconstruction.

Despite the fact that the criteria are similarity categories rather than single concepts, it will still be possible to outline some general relationships between them. There are two particular results I will focus on: (i) the role of the only dynamical criterion, determinism, to conceptually demarcate chaos from randomness and other similar concepts; (ii) the relationship of the concept of unpredictability to the different criteria and the possible use of this concept as a defining criterion itself.

The decomposition of chaos definitions into the five criteria also leads to a natural tidying up of vocabulary: many of the different concepts encountered in chaos theory can now be identified as embodiments of a given criterion for chaos and can then be grouped accordingly (Table 3.1).

3.3.1   Determinism: A Dynamical (Pre-)criterion

As we have seen in section 3.3.1, both May and Osler (1976) and Devaney (1989) emphasize the fact that the models they are investigating are deterministic ones. Similar passages can be found in the large majority of studies on chaos (e.g. Tsonis, 1992, p. 239; Cvitanovic, 1986, p. 3; Ott et al., 1994, p. 13). The meaning of "deterministic" is here sharply distinguished from the concept of "predictability", which will be discussed in section 3.3.4 and section 3.3.5. In fact, determinism in this context is a purely dynamical concept: it refers to the fact that the equations comprising the canonical description of the dynamics of a given model admit a unique solution for a given initial values and therefor do not contain any probabilistic terms.

Indeterministic Models
In view of a fundamental difficulty in the evaluation of chaotic models, which I will discuss in Section 4.3.2, it will be worthwhile to spend some time considering what this conception of determinism implies. For this, it is useful to consider a canonical indeterministic model as a contrasting example. One such model, which, in section 3.2.2, was indeed used as an exemplar of randomness by May and Osler (1976), is the model of a coin toss. The coin toss model is usually used as the exemplar of a model of a Bernoulli process. We recall: a Bernoulli process could roughly be defined as a sequence of trials with a given set of outcomes whereby the probability for a given outcome to instantiate remains the same during each trial. A somewhat more general definition of a Bernoulli process is as a process whose dynamics can be described by a uniform probability function on the space of possible outcomes. In the case of the coin toss, this is usually formulated as the requirement that all possible sequences generated by an infinite series of trials (tosses) have the same probability of instantiating (e.g. Werndl, 2009c). The precise technical definition of this notion of randomness requires considerable measure theoretical background assumptions.

In practice, descriptions of the dynamics of the coin toss usually focus on a single trail in a possible sequence. The dynamics of the coin toss model are therefore usually given as

$$p(T) = p(H) = \frac{1}{2}, \tag{3.1}$$

where $p(H)$ is the probability of throwing heads and $p(T)$ is the probability of throwing tails. Underlying this equation is a suitable definition of the term "probability" and its accompanying calculus, which, in the case of the coin toss, we can take to be the standard probability calculus.

In chaos theory, another important comparison class of models are deterministic models with a probabilistically determined error source, for example, models whose dynamics $F(x)$ has of both a deterministic part $f(x)$ as well as a probabilistic part $p(x)$:

$$F(x) = f(x) + \epsilon(x), \qquad (3.2)$$

where

$$p(\epsilon(x)) = p(x), \qquad (3.3)$$

that is the value of the "noise" is determined through a probabilistic function. In the case of noise, $p(x)$ is often assumed to have a uniform probability distribution, that is to be stochastic under a definition roughly similar to that based on the Bernoulli process as an exemplar.

In the indeterministic models discussed here, the probabilistic formalism has been chosen for practical reasons. A fully deterministic description of the processes in question would be possible in principle (in contrast, e. g. to the case of a quantum process). However, these processes involve dynamics which are difficult to quantify exactly: for example, in the case of the coin toss, the processes responsible span different spatial scales, which makes a fully integrated description impossible; in the case of a noise function, the different mechanism behind this function might simply be unknown or too variable to quantify precisely. Roughly, an indeterministic model can therefore be viewed as a model that corresponds to a model system with dynamics that are not quantifiable exactly, for which a probabilistic description (e.g. that of a Bernoulli process) has been accepted as canonical.

The notions of both indeterminism as well as probability and, in particular, the anthropocentric component in their definitions (introduced by the dependence of their conceptualizations on available information about a process) are subjects of ongoing philosophical debates (e.g. Earman, 1986; Hájek, 2012) and clearly merit more attention than I have been able to give them in this section. Since I am only concerned with indeterministic models as a comparison case to the deterministic models used in chaos theory, my analysis will not go beyond this brief sketch

Deterministic Models

We can now develop a minimal notion of determinism as a property of models that is suitable to distinguish the class of deterministic models from the class of indeterministic models as described above. Namely, the criterion requires that a deterministic model's equations, which are accepted formalization of the model's dynamics, uniquely determine the behaviour of a model for each given initial value. Accordingly, the space of possible outcome for a deterministic model is not best described by a uniform probability function.

The formalization of a deterministic models usually does not contain any probabilistic terms. This implies that the dynamics of the model are viewed as exactly quantifiable on the desired level. It is notably that dynamics here refers to the state of the model after the accepted simplification and idealizations have been made. This could be rephrased more precisely as requiring that the processes corresponding to the model system have exactly quantifiable dynamics.

All of the case studies introduced in Sections 2.3 and 2.4 fulfil such a criterion of determinism: their dynamics are fully quantifiable and their equations contain no probabilistic parts. It is also easy to verify that determinism, in the sense outlined above, is implicitly assumed for virtually all models in chaos theory (e.g. Tsonis, 1992, p. 239; Cvitanovic, 1986, p. 3; Ott et al., 1994, p. 13).

Determinism as Demarcation Criterion

In section 3.1, I have explained that the use of determinism as a defining criterion is unusual since dynamical criteria seldom feature in definitions of descriptive labels like "chaos". However, the case study of the diagnosis of chaos in the discrete logistic model (section 3.2.2) illustrates why the use of this criterion is necessary to distinguish chaos from other similar concepts. In particular, it demarcates chaos from "true stochasticity", that is, the label given to the behaviour of a Bernoulli process. The definition of chaos used by May and Osler (1976) to describe the discrete logistic model (Section 3.2.2), namely stochastic chaos (Section 3.4.4), stipulates that the model's phenomenology is indistinguishable from the phenomenology of a truly stochastic model. The need for an additional dynamical criterion to distinguish between the two labels is therefore immediately apparent. Werndl (2009a) also shows that the behaviour of many models with SDIC (Section 3.3.5), which is another prevalent phenomenological criterion for chaos, can equivalently be generated by deterministic or stochastic dynamics. In this study, her notion of equivalence is a formal, measure theoretical one and her notion of stochasticity is that exemplified by a Bernoulli process as discussed above. The fact that the phenomenological criteria for chaos (which I will discuss in detail in Section 3.3.2 – 3.3.5) are not sufficient to establish chaos as an independent concept is therefore also supported by formal results.

Since the requirement of determinism is needed to define the concept of chaos itself, I therefore maintain that determinism is best interpreted as a criterion for the diagnosis of chaos rather than a sufficient condition for the display of chaotic behaviour. The use of this criterion is one of the most idiosyncratic features of chaos theory. We will see in Section 4.3.2 that determining whether a natural system is chaotic, that is, fulfils the dynamical criterion of determinism, or stochastic, that is should be described by probabilistic dynamics, usually requires a consideration of the behaviour of the system in parameter regimes other than the chaotic phase. This necessity also follows from the phenomenological indistinguishability of some chaotic and stochastic systems, of course.

3.3.2   Transitivity

In contrast to the dynamical criterion of determinism discussed above, the criterion of transitivity is a phenomenological one, that is it puts demands on the behaviour of a model or system rather than on its underlying equations or mechanism. The embodiments of the criterion of transitivity fall into two main classes: topological and statistical formalizations.

Topological Embodiments
An example of a topological embodiment of transitivity, which is particularly applicable to continuous maps like the iterated logistic model, is given by Definition 4 (Section 2.3.2): this definition requires that points from one arbitrarily small neighbourhood will eventually be mapped into any other arbitrarily small neighbourhood. The pre-formal idea behind this concept is an eventual spreading of a system's state throughout a particular region of an appropriate phase space. In Section 2.4.2, we have encountered another topological embodiment of the criterion, which fitted the iterated Lorenz model: there, transitivity was defined as the existence of least one trajectory which visits all possible subregions of an appropriate region of phase space.

Statistical Embodiments

The most prevalent example of a statistical embodiment of transitivity in the concept of ergodicity, which (Berger, 2001, p. 92) describes as the "statistical analogue of topological transitivity". Tsonis (1992, p. 44) gives a formal but non-technical definition of ergodicity:

> [A] system is called ergodic if a long time average of the system is equal to ensemble averages of many different systems...An ensemble is a great number of systems of the same nature but differing in configurations and velocities at a given instance such that they embrace every conceivable combination of configurations and velocities at a given instance.

The specification of "configurations" and "velocities" shows that this description is geared towards mechanical systems. However, these terms can be substituted by generic specifications of microstates. For example, one could prepare an ensemble of discrete logistic models by varying the initial population size N_0 (Section 2.3.2). This embodiment of transitivity therefore statistically captures the same pre-formal idea, namely the fact that the system will "make use" of all of the available phase space.

All embodiments of transitivity require the definition of a region of phase space on which the model is transitive, that is which it eventually spreads through. Accordingly, embodiments of transitivity may not only differ by their choice of topological or stochastic formalization but also by their choice of a relevant phase space region. In chaos theory, the specified region is often an attractor (Sections 2.3.3 and 2.4.1) rather than the whole of phase space. In these cases, the criterion is often formulated as a requirement of the existence of a region in phase space on which the model's behaviour fulfils an embodiment of transitivity rather than as a requirement of transitivity as such.

All Bernoulli processes are ergodic (e.g. Werndl, 2009a). Accordingly, stochastically chaotic models, which are by definition phenomenologically indistinguishable from Bernoulli processes, are also ergodic. However, in the definition of stochastic chaos (Section 3.4.4), this criterion is not explicitly required. In contrast, in the definition of Devaney chaos (Definition 5, Section 3.4.1), an embodiment of transitivity is explicitly required. Combined with the requirement of a dense set of periodic points, it implies a third criterion, SDIC (Section 2.3.3; Banks et al.,1992). A third definition of chaos, mixing (Section 3.4.1), is explicitly designed to imply both ergodicity and an embodiment of SDIC. Accordingly, the relationship of the criterion of transitivity with other criteria in a given chaos definition varies depending on the combination of criteria and their embodiments used in the definition.

3.3.3   Periodicity

The criterion of periodicity is a phenomenological one. There are two broad classes of embodiments of this criterion: strong and weak embodiments.

Strong Embodiments
In Section 3.2.3, we have seen that the criterion of periodicity is part of the definition of Devaney chaos (Definition 5, Section 3.4.1), namely as the requirement of a dense set of periodic points. As we have seen in Section 2.3.3, in the iterated logistic model, this implies that after an infinite number of iterations, the set Λ will consist of an infinite number of fixed points and their limit points. The model therefore fulfils this embodiment of the criterion. Similarly, the iterated Lorenz model possesses a dense set of periodic points (Section 2.4.2).

This form of the criterion of periodicity is formally strong and, in particular when viewed in light of the coexisting criterion of aperiodicity (Section 3.3.4), it seems to contrast with some of the crucial pre-formal ideas about chaos. Smith (1998, p. 176) calls the requirement "the Devaney's definition somewhat surprising emphasis on periodic orbits" and suggests that this emphasis disqualifies the definition of Devaney chaos from becoming canonical. Smith (1998, p. 176) also constructs an example of a toy-model that would possess high periodicity in the sense of the Devaney definition, but would not be aperiodic. I agree with Smith (1998) that definitions of chaos that are based on strong embodiments of periodicity and aperiodicity, respectively, will probably not have completely overlapping extensions. However, given the fact that different chaos definitions are usually designed to be particularly applicable to a given class of models, a complete overlap in extensions should not be expected. Furthermore, the periodic properties highlighted by the definition of Devaney chaos in horizontal models like the iterated logistic or Lorenz model (Sections 2.3.3 and 2.3.2) are precisely those characteristics that are also used the investigation of the aperiodic properties of the discrete logistic and Lorenz models. Therefore, the conceptual differences in the different chaos definitions, including the use of embodiments of prima facie contrasting criteria of periodicity and aperiodicity, in many cases seem to be justified by the applicability of those definitions to different classes of models and their usefulness in highlighting the specific properties of those models that will be crucial to their investigative role.

Weak Embodiments

There exist several weaker embodiments of the criterion of periodicity. Among these, I count requirements of the existence of some regular structure (e.g. Berger, 2001, p. 40). Often, this structure is taken to be an attractor, that is the existence of a finite region in phase space to which the behaviour of the system is eventually confined (Section 3.2.2). In a related embodiment, the attractor is additionally required to be fractal or of fractal dimension (e.g. Smith, 1998). The mere existence of an attractor would not be sufficient to render the behaviour of chaotic systems phenomenologically novel; many deterministic systems possess attractors (e.g. the simple harmonic oscillator attractor being an elliptic region in phase space). Therefore, this criterion is usually combined with other criteria for chaos, which the system should fulfil on the attractor, most prevalently SDIC (Section 3.3.5) or transitivity (Section 3.3.2). Attractors that fulfil additional requirements of chaos are often called "strange attractors". The definition of chaos through the requirement of the existence of a strange attractor will be discussed in Section 3.4.5.

3.3.4   Aperiodicity

The phenomenological criterion of aperiodicity is used by May and Osler (1976) to describe the chaotic behaviour of the discrete logistic model (Section 3.2.2). The criterion is a phenomenological criterion and its various embodiments consist of the specification and formalizations of properties that are considered characteristic of either (i) aperiodic behaviour itself or (ii) the phenomenology of truly stochastic systems as defined in Section 3.3.1. There is also a form of unpredictability associated with the criterion of aperiodicity, which can also be used as an embodiment of the criterion.

Absence of Periodicity

A relatively straightforward way of defining aperiodicity is by the requirement of a uniform Fourier spectrum, that is by requiring that that no predominant frequencies can be established in a Fourier analysis of the relevant data (e.g. for review, Abarbanel et al., 1993). This embodiment of the criterion is particularly applicable to the analysis of empirical data sets and of the output of numerical models. Similar embodiments of a criterion establishing the absence of patterns in the output of model are measure of entropy, that is Kolmogorov entropy (e.g. Batterman, 1993) or low correlation properties (e.g. Ott et al., 1994, Chapter 10). Since the definition of aperiodicity is itself a debated concept in philosophy and mathematics (e.g. Earman, 1986, Chapter 8; Smith, 1998, Chapter 9), the choice of an embodiment of aperiodicity can also be influenced by the author's viewpoint on this debate. For example, there is still disagreement on the question of whether aperiodicity should be seen as a concept fundamentally associated with a lack of recognizable patterns or with a lack of relevant information.

Apparent Stochasticity
In chaos theory, the most prominent class of embodiments of the criterion of aperiodicity are those that aim to establish apparent stochasticity, that is to show that the behaviour of a chaotic model is indistinguishable from the behaviour of a Bernoulli process. In order to establish indistinguishability, suitable characteristics of the phenomenology of a Bernoulli process need to be identified. As mentioned in Section 3.3.1, during a Bernoulli process, the probability for obtain a given outcome in each trial in a sequence of trials remains the same (by definition). In the case of the coin toss, this implies that for each single coin toss the probabilities of obtaining heads or tails is always $1/2$. Consequently, the outcomes that have been previously obtained in a sequence of trials cannot have any bearing on the outcome of the next trial, that is the outcome of given coin toss is independent of the history of previous coin tosses (e.g. Werndl, 2009b). This can be interpreted to imply that no discernible patterns can exist in such sequence. Accordingly, the embodiments of aperiodicity described above can also be used as embodiments of apparent stochasticity if they are seen as phenomenological features that capture the behaviour characteristics of a Bernoulli process best.

As described in Section 3.3.2, all Bernoulli processes are ergodic. Ergodicity, which we have already encountered as an embodiment of transitivity, has therefore also been used as a marker for apparent stochasticity (e.g. Lauwerier, 1991, p. 91). We recall: a model is ergodic, if the average behaviour of the model over a suitably long integration time equals the average of an ensemble of appropriately prepared models, for example, of models with slightly altered initial conditions. An embodiment of the criterion of aperiodicity through ergodicity is therefore a statistical formalization.

The specification of formal properties that a chaotic model should share with a truly stochastic model in order to be viewed as indistinguishable from it is complicated by the fact that the formalization of the phenomenological manifestations of stochasticity is itself an unresolved issue. For example, while all Bernoulli processes are ergodic, the relationship between periodicity and the statistical property of ergodicity is itself a topic of ongoing research (Berger, 2001, p. 121) and usually requires a careful technical definition of a suitable probability and phase-space measure. Using ergodicity as an embodiment of aperiodicity will therefore capture an aspect of indistinguishability from a truly stochastic system. But it might not ensure that the system under investigation is aperiodic according to other criteria, or that there are no non-stochastic and non-chaotic systems which are also ergodic.

Local Unpredictability
There is a form of unpredictability that is closely related to the absence of patterns in the behaviour of an apparently stochastic system or model. I will name the unpredictability associated with apparent stochasticity as "local unpredictability" to distinguish it from another, "global" form of unpredictability associated with the criterion of SDIC (Section 3.3.5). Since past events in the behavioural history of an apparently stochastic model, that is of a model whose phenomenology is indistinguishable of that of a Bernoulli process, are statistically irrelevant to the future development of the model, it is not possible to predict the behaviour of the model from a given point in its development. In the standard example for a Bernoulli process, the coin toss, it is not possible to predict the outcome of the next coin-toss in a sequence from earlier tosses. Similarly, it is apparent in Fig. 2.4 that the behaviour of the discrete logistic model will be unpredictable in the same local sense.

Local unpredictability as described here is, of course, a direct consequence of the independence of each new event from the behavioural history of a system and the corresponding lack of behavioural patterns associated with apparent stochasticity. I therefore agree with Batterman (1993) that local (and, as I will argue in Section 3.3.5, global) unpredictability is not a defining criterion of chaos (as proposed by, e.g., Stone, 1989, p. 123). Instead, as described above, the concept forms the conceptual core of some embodiments of the criterion of aperiodicity, namely those that are based on testing the possibility of extrapolation from a given set of data rather than on analysing its periodicity or ergodicity.

As noted by Smith, (1998, chapter 10), the criterion of aperiodicity can be seen as conflicting with the criterion of periodicity. However, as discussed in Section 3.3.3, this contrast is strongly mitigated by the facts that (i) specific embodiments of the two criteria can be compatible with each other and (ii) the models targeted by given embodiments of the criterion are often not the same. The majority of the embodiments of the criterion of aperiodicity appear to be geared towards a description of the behaviour of discrete models, for example, the discrete logistic model (Section 2.3.2). In contrast, the majority of the embodiments of the criterion of periodicity (Section 3.3.3) appear to be geared towards a description of the behaviour of iterations of continuous functions, that is the iterated logistic and Lorenz models (Sections 2.3.3 and 2.4.2). The case of a model and two definitions of chaos, which are both directly applicable to the model and require strong embodiments of periodicity and aperiodicity, respectively, appears to be rare in practice (although Smith [1998, p. 176] constructs such a situation). Furthermore, the highlighting of different properties in different models in a given lineage through the use of different definitions of chaos seems to serve an epistemic purpose: it highlights those properties that will be important in the later evaluation of those models (Chapter 4).

3.3.5   Sensitive Dependence on Initial Conditions

The criterion of SDIC was first introduced in Section 3.3.2, where it appeared as a consequence of the criterion of aperiodicity in the definition of stochastic chaos. In the definition of Devaney chaos (Section 3.2.3), a specific embodiment of the criterion, given by Definition 4, is required. As explained in Section 2.3.3, the definition is based on the notion that points or trajectories that are initially very close together will eventually be separated by a relatively large distance. The criterion of SDIC is a phenomenological criterion, that is it pertains to and can be diagnosed from the model's behaviour. This might not be immediately obvious in this case since establishing SDIC usually requires knowledge about several different trajectories of a model or system. However, while the generation of these trajectories requires the repeated running of a given model, in principle, no detailed knowledge about the dynamics of the model is necessary. The fact that such repeated runs are usually not available for natural systems, of course, presents a technical difficulty in the diagnosis of SDIC for natural systems. However, this is also solved by a particular mode of analysis of empirical data (Section 4.3.2).

The straightforward meaning of the term "initial conditions" is the set of initial values of the dependent variables, for example, the initial population value N_0 in the discrete logistic model (Section 2.3.2) or the set of initial values (X_0, Y_0, Z_0) for the dependent variables in the discrete Lorenz model. However, Mayo-Wilson (2015) pointed out correctly that many models contain other numerical constants, which are also specified prior to the integration of the model, for example, the growth rate r in the discrete logistic model and the parameters σ, b and r in the discrete Lorenz model. Chaotic models often display sensitivity to changes in both sets of values. In this section, I will primarily discuss the former kind of SDIC, that is sensitivity to changes in the numerical values of dependent variables. However, the results of the discussion should be directly applicable to the second form of SDIC as well.

There is another aspect of the meaning of "initial conditions" that warrants further thought. The concept of SDIC as described above aims to capture the influence that small changes in the values of the dependent variables at the very beginning of the model's development will have. However, if the development of the model is followed for a long enough time span, that is if it is integrated or iterated for a sufficiently large number of steps, then any measure of SDIC is also a measure of how changes at intermediate times of values of the dependent variables will influence the remainder of the model's development. For example, in the iterated logistic model (Section 2.3.3), two neighbouring points $N^{(n)}(x_1)$ and $N^{(n)}(x_2)$ on the polynomial function generated by the nth iteration will, after a suitable number of k further iterations, be separated by the same minimal distance by which two neighbouring points $N^{(0)}(x_1)$ and $N^{(0)}(x_2)$ on the initial line function are separated after the same number of k iterations.

In the case of the iterated logistic model, this property appears to be prima facie a trivial consequence of the need to fix an arbitrary reference value for the dependent variable. However, this "ongoing" kind of SDIC assumes a different importance in the case of discrete models, and in particular in the case of numerical models (e.g. like the discrete Lorenz model, Section 2.4.1). For these kinds of models, it has been hypothesized that the truncation errors introduced into the model at each step of integration will then also be subject to SDIC, leading to an ongoing distortion of the solution of the model (e.g. for an illustration of this amplification of errors Schurz, 1996, p. 128). As I will discuss below, this effect has been associated with another form of unpredictability of chaotic models, viz. global unpredictability, which has also been used in a particular definition of chaos.

Embodiments of the criterion of SDIC appear to divide into two large classes: those very similar to the topological Definition 4 (Section 2.3.3) and those that require an exponential divergence of trajectories. The latter class of embodiments will be discussed in the next paragraph.

Exponential Divergence

In embodiments requiring exponential divergence, the distance δ in Definition 4 is replaced by an exponential function of the model's independent variable $\delta(x) = \delta_0 e^{\lambda(x)x}$. The exponent $\lambda(x)$ is called the Lyapunov exponent. It is easy to see that, if the Lyapunov exponent is positive, the distance between the two points will exponentially increase as the model evolves. Allowing $\lambda(x)$ to be a function of x itself means that the increase can be more complicated than trivially exponential (e.g. for a more detailed introduction to Lyapunov exponents, Alligood et al., 1997, pp. 108–109).

There are technical variations on how the Lyapunov exponents are computed for a given model or data set and which precise requirements are put on it (e.g. for review, Abarbanel et al., 1993). While most of the embodiments are applicable to models with continuous trajectories, it is also possible to compute Lyapunov exponents for discrete models. Furthermore, there are variations in the embodiments with respect to the requirements on the exponent itself. However, in order for a model to be diagnosed with SDIC it is usually required that $\lambda(x)$ remains positive over an appropriate length of time or other variable range.

SDIC and Other Chaos Criteria

In Section 3.2, we have already encountered two examples in which embodiments of SDIC were shown to be implications of other criteria of chaos. May and Osler (1976) describe SDIC as a consequence of apparent stochasticity, that is as a consequence of an embodiment of the criterion of aperiodicity (Section 3.2.2). Heuristically, the argument behind this notion rests on the fact that small differences in the initial conditions will be amplified through the stochastic fluctuations in the model's behaviour (Fig. 2.1), that is a small difference in initial condition might determine whether at time step n the model will produce a downward or upward spike, hence rendering these two states very different from each other. Formally, it could be shown that SDIC is a consequence of the equiprobability of outcomes of a Bernoulli process (Section 3.3.1).

As described in Section 3.2.3, Banks et al. (1992) proves that the embodiment of SDIC required by Devaney chaos (Definition 4) is redundant, that is the two other embodiments of transitivity and periodicity in Definition 5 imply the existence of this embodiment of SDIC. Accordingly, for the iterated continuous functions that the definition is applicable to, SDIC is a consequence of the combination of two other chaos criteria.

While the relevant combinations of other criteria in these scenarios can be viewed as sufficient conditions for SDIC, this analysis remains on a purely phenomenological level: none of these accounts about why a model displaying some other chaotic properties will also display SDIC requires any analysis of the model's dynamics. In other words, these accounts merely state that if the model has certain phenomenological criteria, then it will also be possible to diagnose the model as having the given embodiment of SDIC. These relationships between different criteria therefore still pertain to the description of chaotic behaviour rather than the analysis of the sufficient conditions for the occurrence of chaos, which will be discussed in Chapter 4.

Global Unpredictability

Like the criterion of aperiodicity (Section 3.3.4), SDIC has also been associated with a particular notion of unpredictability. However, I maintain that the concept of unpredictability associated with SDIC is subtly but importantly different from the notion of unpredictability associated with aperiodicity. A similar distinction between two kinds of unpredictability can also be found in Werndl (2009b). We recall: I used the term "local unpredictability" to describe our inability of extrapolation from previous events, which is associated with aperiodic behaviour. In contrast, SDIC has been associated with a kind of unpredictability, for which I will use the moniker "global". An equally apt description would be "practical". Models with SDIC are globally unpredictable because SDIC implies that small errors in the initial conditions – or in intermediate values – of a model will lead to very different outcomes. Since there is a practical limit on the precision with which initial conditions can be specified, the range of possible results resulting from a given set of initial conditions will be significantly larger than that of comparable models without SDIC. Accordingly, the predictive powers of models with SDIC are significantly lower than those without SDIC.

Since this type of unpredictability is only associated with SDIC, other criteria for chaos do not need to be present for a model to be globally unpredictable: for example, any model whose trajectories separate exponentially in response to small changes in the initial conditions, and which therefore has positive Lyapunov exponents, will suffer from global unpredictability. However, since SDIC is implied by a number of (combinations of) other criteria of chaos, global unpredictability can also be a consequence of combinations of these criteria. In particular, since SDIC is implied by the majority of embodiments of aperiodicity, aperiodic models are both locally and globally unpredictable.

The implications of global unpredictability for scientific practice, in particular, for forecasting with models that might possess SDIC, have been subjects of wide and ongoing debate between scientists and philosophers (e.g. from a philosophical viewpoint, Stone, 1989; Smith, 1998, chapter 4; Werndl, 2009b; from a technical viewpoint, Judd and Smith, 2001; Judd and Smith, 2004). I do not wish to enter this debate. However, here and in Section 4.3, I would briefly like to discuss attempts to define chaos as the existence of global unpredictability in a model. In the next paragraph, I will argue (i) that global unpredictability should be seen as a consequence of other criteria for chaos rather than as a defining criterion itself; and (ii) that definitions based on global unpredictability as such have a very narrow range of applicability, namely, they seem to apply mostly to numerical models. Then, in Section 4.3, I will argue that such definitions also have philosophically undesirable consequence for the positing of sufficient conditions for the occurrence of chaos.

Global Unpredictability as a Defining Criterion of Chaos

There are a few discussions of the definition of chaos that propose to posit global unpredictability as at least one, if not the only, criterion for chaos (e.g. Chernov and Markarian, 2006, pp. 48–49; Schurz, 1996, p. 127–130). In some of these cases, no distinction between global and local unpredictability (Section 3.3.4) is made, so that it is not always clear whether the proposed criterion is not actually an embodiment of aperiodicity. However, in the two accounts cited here, it seems clear from the context of the discussion that the authors mean the kind of unpredictability associated with the magnification of uncertainties in the initial or intermediate values of a model through SDIC. For example, Chernov and Markarian (2006, pp. 48–49) provide a two-page discussion of the exponential amplification of truncation errors in a numerical model with SDIC (in the embodiment of positive Lyapunov coefficients) and then conclude that:

> The implications of these fact are far reaching and can be characterized by the unifying term, chaos. The most important feature of a chaotic system is that its distant future and remote past become independent of its present state and can be describe only on, "on average", in probabilistic terms.

While such a description would fit the concept of local unpredictability as well, it is apparent from the context here that the development of the model becomes unpredictable due to the fact that the approximation provided by the numerical model becomes exponentially worse with integration time.

My first criticism of attempts to use global unpredictability as a defining criterion of chaos is similar to that made by Werndl (2009b): proposing global unpredictability as a defining criterion somewhat obfuscates the fact that this property is implied by other criteria of chaos. As a matter of fact, most authors who propose such definitions of chaos seem to assume that the model will fulfil embodiments of the criterion of aperiodicity in addition to being globally unpredictable (e.g. Chernov and Markarian, 2006, pp. 48–49; Schurz, 1996, p. 140). In view of the formal results on the implication of SDIC by other criteria for chaos discussed above, it seems that a definition of chaos based on these criteria would better serve the purpose of maximum clarity.

There is a second problematic aspect to using global unpredictability as a defining criterion for chaos: in order for a model to become globally unpredictable in principle, one also needs to assume that there exists an in-principle limit of precision or an in-principle error source to supply such small imprecisions. One class of models with such an error source are numerical models, which are subject to the computer's limit of precision. Accordingly, a definition of chaos that heavily relies on global unpredictability as a criterion seems to be applicable mostly to these models. For example, it is difficult to see how such a definition of chaos would apply to the iterated Lorenz or logistic model (Sections 2.3.3 and 2.4.2). This would imply that chaos becomes primarily a property of numerical models and could be viewed as a measure of the inaccuracy of these models. However, as I will briefly discuss in Section 4.3, viewing chaos as a property of numerical models leads to (in my opinion) unresolvable problems in simultaneously construing chaos as a property of natural systems and in determining sufficient conditions for the occurrence of chaos. I will also argue that there is a better interpretation of how numerical models are usually treated in chaos theory: namely, that much investigative effort is spent on ruling out error propagation and other effects resulting from the discrete nature of these models. In Section 4.5, I will illustrate this treatment through the case study of the evaluation of the lineage of Lorenz models.

3.4   ANALYSIS AND COMPARISON OF CHAOS DEFINITIONS

In this section, I will demonstrate how the five criteria identified in Section 3.3 can be used to analyse different definitions of chaos. Five definitions of chaos will be discussed in detail: Devaney chaos (Section 3.4.1); (the definition of chaos as) mixing (Section 3.4.2); (the definition of chaos in terms of) positive Lyapunov exponents (Section 3.4.3); stochastic chaos (Section 3.4.4) and (the definition of chaos in terms of) strange attractors (Section 3.4.5). Each of these definitions will be shown to require different combinations and embodiments of the five criteria for chaos. While it would be impossible to compile an exhaustive list of chaos definitions, these five definitions appear to cover the predominant combinations of criteria we see in chaos theory. I will also give a brief overview of the models which have been diagnosed as chaotic according to each definitions, that is I will provide a brief description of each definition's extension.

As described in Section 3.3, my framework of analysis is based on a view of chaos definitions as being doubly decomposable into (i) combinations of general criteria and (ii) their formal embodiments. This section will demonstrate how the decomposition of chaos definitions into these general criteria and embodiments can be a used as a heuristic tool in the analysis and comparison of different chaos definitions. Due to the fact that the criteria are similarity categories of differently formalized concepts rather than single, uniquely defined concepts, this analysis is not exact.

Table 3.2 Five prevalent chaos definitions and their criteria for diagnosing chaos

	Determinism	*Periodicity*	*Transitivity*	*SDIC*	*Aperiodicity*
Devaney C.	X	X	X	(X)	–
Mixing	X	–	X	X	–
Lyapunov exp.	X	–	–	X	–
Stochastic C.	X	–	(X)	(X)	X
Strange attr.	X	X	[X]	X	[X]

Note: If a criterion is implied by other criteria but is not explicitly stated in the definition, it is shown in round brackets. If a criterion is only occasionally required, it is shown in square brackets

However, I will be able to highlight three interesting features of the use of definitions in chaos theory: (i) that all chaos definitions require the dynamical (pre-)criterion of determinism in addition to variable combinations of phenomenological criteria; (ii) that there are many overlaps in the extensions of different chaos definitions; and (iii) that some embodiments are used to specifically gear a definition towards a particular class of models while other embodiments have been chosen with the aim of encompassing as many models as possible. As described in Section 3.3, the use of different embodiments and combinations of criteria for in chaos is therefore often deliberate and guided by the investigative purpose of a particular class of models.

The five definitions and the criteria they can be decomposed into are summarized in Table 3.2.

3.4.1   Devaney Chaos

The formal definition of Devaney chaos is given in Definition 5. As is apparent in Definition 5, Devaney chaos requires three phenomenological criteria for the diagnosis of chaos: periodicity (Section 3.3.3), transitivity (Section 3.3.2) and SDIC, whereby the latter is implied by the last two criteria (Section 3.3.5). All of these criteria appear in formal, topological embodiments (Section 2.3.3), which are best suited for the description of iterated continuous maps. The restriction of the applicability of the definition to these maps (Definition 5), means that the definition also implicitly requires the dynamical criterion of determinism (Section 3.3.1).

As discussed in Section 3.3.3 and highlighted by Smith (1998, Chapter 10), Devaney chaos requires a strong embodiment of periodicity. This means that there are some models that are chaotic according to other chaos definitions but will not fulfil Definition 5. In particular, models like the discrete logistic model (Section 2.3.2) or the toy model constructed by Smith (1998, p. 176), which are strongly aperiodic, will not be judged Devaney chaotic. However, since these discrete models fall outside the primary domain of applicability of Definition 5, this does not lead to a direct conceptual conflict in the classification of these models. Furthermore, in Section 4.2.2, it will become apparent that the definition highlights precisely those properties of the iterated logistic model, which are important for the investigative use of the model within the lineage of logistic models. The extension of Devaney chaos also overlaps with a number of other chaos definitions, in particular with those of chaos as mixing (Section 3.4.2); the definition of chaos in terms of positive Lyapunov coefficients (Section 3.4.3) and the definition of chaos in terms of strange attractors (Section 3.4.5). This is explained by a corresponding overlap in the criteria required by these definitions (Section 3.2).

Extension of Devaney Chaos
Definition 5 formally applies to continuous, iterated maps like the iterated logistic model (Section 2.3.3) and the iterated Lorenz model (Section 2.4.2). Both of these models are Devaney chaotic (Section 3.2.2; Hirsch et al., 2004, Chapter 14). The discrete Lorenz model has been indirectly diagnosed with Devaney chaos: attempts to show that the dynamics of this numerical model correspond to those of the iterated Lorenz model will be discussed in Section 4.5. Other well-known one-dimensional iterated models that are Devaney chaotic include $N(x) = \tan x$, the baker map and the tent map (Devaney, 1989, p. 52).

Some "billiard balls scenarios" have been found to be Devaney chaotic. Most models of billiards assume that the billiard table has a rigid boundary of some shape, an elastic ball and a frictionless surface. Billiard models are therefore at heart vertical models (Section 2.2.1), which derive from the governing theory of classical mechanics and a prepared description of the given billiard system. However, often a long lineage of horizontal models is spun off from a given vertical billiard model: for example, by running through a variation of boundary shapes, the choice of which is usually purely mathematically motivated (e.g. model construction in billiard dynamics, Berger, 2001; Chernov and Markarian, 2006). While the properties of many billiard scenarios are not yet well understood (Chernov and Markarian, 2006, p. ix), it has been shown that models of billiards on rationally polygonal (Boshernitzan et al., 1998) and curved convex tables (Berger, 2001; Section 2.6) possess at least very similar forms of periodicity, transitivity and SDIC as required by Definition 5.

3.4.2 Mixing

Werndl (2009b, p. 204) proposes to call a system "chaotic" if it is mixing in some region of its phase space, that is if in this region "any bundle of solutions spreads out in phase space like a drop of ink in a glass of water". Conceptually, this seems to by equivalent to requiring two phenomenological criteria for chaos (Werndl, 2009b, p. 203): transitivity (Section 3.3.2) and SDIC (Section 3.4.5). Note that, in contrast to its redundant role in Devaney chaos (Section 3.4.1), SDIC is required here as an explicit criterion. Werndl (2009b, p. 204) then casts these two criteria into one embodiment, the measure-theoretical concept of mixing, which implies measure-theoretical notions of mixing, in the embodiment of ergodicity (Section 3.3.2) and of SDIC (Werndl, 2009b, pp. 205–206). The formal definition of mixing requires background knowledge in measure theory; I will therefore not display it here. However, conceptually, the definition of mixing proposed by Werndl (2009b) requires that the measure of the distance between the images of any two subregions of a model's phase space becomes independent eventually of the measure of the initial distance between these two regions, that is whether two regions were initially located close together is no predictor for where these regions will eventually be mapped to. Werndl (2009b, p. 197) also states that mixing systems must be deterministic.

The definition of chaos as mixing has been deliberately designed with universality in mind (Werndl, 2009b, p. 204), that is its extension is meant to include as many of the models which have been judged chaotic in chaos theory as possible (while still maintaining appropriate conceptual boundaries). My analysis here reveals that the definition possesses two features, which make it particularly suited to this purpose: (i) it requires precisely those two criteria for chaos which are also required by four of the five chaos definitions discussed here (Table 3.3); (ii) its measure-theoretical embodiment of these criteria means that the definition is applicable to a wide range of models. In particular, it is directly applicable to both discrete and continuous

Table 3.3   Kinds of chaos found in the logistic and Lorenz models (Sections 2.3 and 2.4)

	Dis. Log. M.	It. Log. M.	Dis. Lor. M.	It. Lor. M.
Devaney C.	–	X	[X]	X
Mixing	X	X	[X]	X
Lyapunov exp.	X	–	X	–
Stochastic C.	X	(X)	(X)	–
Strange attr.	–	(X)	[X]	X

If a diagnosis has been made indirectly, that is only on a restricted domain or through a particular, shared embodiment, the corresponding mark is put in brackets. Diagnoses of chaos for the discrete Lorenz model that rely on a direct transfer of results from the iterated Lorenz model have been put in square brackets

models and systems. The overlap between the extension of mixing and those of the other chaos definitions is therefore particularly large.

Extension of Mixing
As apparent in Table 3.2, the definition of chaos through mixing implicitly requires two of the three phenomenological criteria required by Devaney chaos. The extension of mixing therefore seems to include virtually all models judged to be Devaney chaotic, including the iterated logistic model (Section 2.3.3; Werndl, 2009b, p. 203) and the iterated Lorenz model (Section 2.4.2; Luzzatto et al., 2005). The discrete Lorenz model (Section 2.4.1) has again been diagnosed as being mixing, based on the assumption that its dynamics are sufficiently similar to those of the iterated Lorenz model to transfer results from the latter to the former.

As discussed by Werndl (2009b, pp. 208–209), the extension of mixing also largely overlaps with those of the definition of chaos in terms of strange attractors 3.4.5 and in terms of the requirement of positive Lyapunov exponents (Section 3.4.3), although the existence of positive Lyapunov exponents by itself is neither necessary nor sufficient for a model to be mixing. A model with a strange attractor will be mixing if suitable forms of transitivity and SDIC hold on the attractor. Since apparent stochasticity implies the criteria of SDIC and transitivity, stochastically chaotic models (Section 3.4.4) are also found to be mixing, that is the discrete logistic model (Section 2.3.2) can be diagnosed as mixing on the grounds that it is phenomenologically indistinguishable from a Bernoulli process.

As outlined above, the inclusiveness of mixing as a definition of chaos was intended by Werndl (2009b). Accordingly, the successful design of this definition shows that a definition of chaos whose extension includes all exemplary models in chaos theory (as demanded, e.g., by Smith, 1998, Chapter 10) is possible. In this section, I have been able to highlight that this feat involves identifying the two criteria of chaos displayed by all those models and choosing an embodiment of these criteria that is applicable to both discrete and continuous models.

3.4.3   Positive Lyapunov Exponents

As discussed in Section 3.4.5, one embodiment of the criterion of SDIC is the requirement of positive Lyapunov exponents. We recall: This embodiment implies that trajectories with similar initial conditions diverge exponentially. There are several authors who equate the existence of chaos with having positive Lyapunov exponents (e.g. Mayo-Wilson, 2015; Ott et al., 1994, p. 12). Such definitions of chaos are then based on this single phenomenological criterion of SDIC. However, like the other chaos definitions, the dynamical criterion of determinism is implicitly or explicitly required as well (e.g. Ott et al., 1994, p. 10).

The precise definition of how the Lyapunov exponents are to be computed usually depends on the model under investigation; in many cases the exponents are computed as an average of the divergence of different trajectories. Similarly, the conditions on the exponents can be more or less demanding: it might be required that coefficients are positive over a reasonable interval of integration steps or that they are positive at all times. For the discussion in this book, it is not necessary to provide a detailed discussion of such technical differences. However, the interested reader can find several different approaches to the computation of Lyapunov coefficients in Ott et al. (1994, Chapter 4 and Chapter 8).

*Extension of the Definition of Chaos in Terms of Positive
Lyapunov Exponents*
Allowing for some variations across different embodiments, the definition of chaos in terms of positive Lyapunov exponents can be viewed as requiring only one of the three or two criteria for chaos required by Devaney chaos (Section 3.4.1) and mixing (Section 3.4.2), respectively. It is therefore not surprising that positive Lyapunov exponents can be computed for many models that have been found to be Devaney chaotic

or mixing. However, Werndl (2009b, pp. 205–206) points out that not all mixing systems feature exponential divergence of trajectories and that not all systems with positive Lyapunov coefficients are mixing. For example, positive Lyapunov exponents have been established in the iterated logistic model for some, but not for all, growth rates (e.g. Alligood et al., 1997, pp. 114–124). Similarly, the iterated Lorenz model, as constructed in Section 2.4.2, does not necessarily feature exponentially diverging trajectories (Hirsch et al., 2004, p. 323). (We recall: the precise form of the trajectories is not specified in the model since its construction only involves specifying an attractor.) There is therefore probably no complete overlap between the extensions of these three definitions.

As discussed in Section 3.3.4 and 3.3.5, the criterion of aperiodicity in the embodiment of apparent stochasticity implies a form of SDIC. Therefore, positive Lyapunov coefficients can be computed for many stochastically chaotic models, including the discrete logistic model (e.g. Alligood et al., 1997, pp. 109–110). Positive Lyapunov coefficients have also been computed for a number of numerical models, including the discrete Lorenz model (e.g. Alligood et al., 1997, pp. 366–367) and discrete models of the three-body problem (e.g. Gonczi and Froeschle, 1981).

3.4.4   Stochastic Chaos

As discussed in Section 3.3.2 and 3.3.4, a fourth definition of chaos is stochastic chaos, which was used by May and Osler (1976) in the diagnosis of chaos in the discrete logistic model. An explicit conceptual definition of stochastic chaos is also given by Lorenz (1993, p.4):

Definition 6. I shall use the term chaos to refer collectively to processes of this sort – ones that appear to proceed according to chance even though their behaviour is in fact determined by precise laws.

It is easy to see that this definition is based on two criteria for chaos: the phenomenological criterion of aperiodicity in the embodiment of apparent stochasticity and the dynamical (pre-)criterion of determinism. The use of the latter criterion thereby corresponds to the way it was introduced in Section 3.3.1, that is determinism here requires the existence of a unique solution for each initial value and therefore the absence of any probabilistic terms in a model's equations.

As discussed in Section 3.3.4, there are several embodiments of apparent stochasticity. These are usually requirements that the phenomenology of an apparently stochastic models shares a feature that is also characteristic of the phenomenology of a Bernoulli process, for example, that the model is ergodic, or lacks correlation and local predictability. In some cases, including Lorenz (1993), the underlying conception seems to be one of a holistic comparison, that is the requirement is that in a suitable comparison to the behaviour of a Bernoulli process, usually a coin toss, no significantly different features may be found. Thus, the variety of specific embodiments of apparent stochasticity can be explained by the fact the phenomenological manifestations of stochastic dynamics are themselves still a subject of scientific and philosophical debate (Section 3.3.4).

As described in Section 3.3.4, apparent stochasticity, usually implies a form of SDIC and transitivity. However, in most formulations of stochastic chaos, these two criteria are not explicitly required or discussed (e.g. May, 1976; Lorenz, 1993).

Extension of Stochastic Chaos
The class of models most strongly associated with stochastic chaos are discrete, numerically integrated models, like the discrete logistic model (Section 2.3.2) and the discrete Lorenz model (Section 2.4.1). In the latter case, the diagnosis of stochastic chaos is somewhat dependent on both the representation of the model's behaviour and on the chosen embodiment of the criterion: Thus, the model has been found to be ergodic on the whole domain while embodiments based on measures of aperiodicity and lack of correlation are usually applied to one-dimensional representations, that is to the tracing of a single variable as displayed in Fig. 2.4b. Similarly, the iterated logistic model (Section 2.3.3), can be found to be ergodic if a suitable measure is defined on the disjoint domain Λ and on the whole domain for $r = 4$ (e.g. Werndl, 2009b, p. 205).

Lorenz (1993) provides several examples of simple computer-based models, which are stochastically chaotic according to his holistic definition (my Definition 6), for example, a simply discrete model of a skier in a mogul (bump) field. Berger (2001) and Chernov and Markarian (2006) describe several billiard and pinball scenarios as stochastically chaotic; both set of authors use ergodicity as an embodiment of apparent stochasticity.

Stochastic chaos is also often used in the diagnosis of chaos in empirical data from natural systems. While some such studies rely on formalized means of showing indistinguishability from stochastic behaviour with respect to a given characteristic (Section 3.3), others will simply use visual indistinguishability (e.g. for a collection of experimental studies in which stochastic chaos is detected, Robertson and Combs, 1995). A major difficulty in the detection of stochastic chaos in natural systems is a need to show that the criterion of determinism is fulfilled by the system. In Section 4.3.2, I will discuss this difficulty in more detail.

The definition of stochastic chaos is based on a phenomenological criterion – aperiodicity – which has not explicitly featured in the other chaos definitions. While this can be interpreted as a conceptual difference in the way the concept of "chaos" is construed, and models can be conceived that will fulfil this definition but not others (e.g. Smith, 1998, Chapter 10), in practice, this difference is mitigated by the fact that some embodiments of apparent stochasticity are also embodiments of other criteria for chaos. This is particularly true for ergodicity, which is also often used as an embodiment of transitivity. Therefore, a variable construction of which features are characteristic of apparent stochasticity means that the concept is in many cases not clearly divided from embodiments of other criteria for chaos. A second factor mitigating the demarcating power of aperiodicity as a criterion for chaos is the fact that most chaos definitions are domain specific: accordingly, a particular embodiment of apparent stochasticity might be fulfilled by a particular aspect or representation of a model that does not directly seem to fulfil Definition 6.

Since SDIC is part of the definition of stochastic chaos, there is also a large overlap between the extensions of a definition of chaos through the requirement of positive Lyapunov exponents (Section 3.4.3) and that of stochastic chaos. However, while some form of exponential divergence can be found in many stochastically chaotic models, e. g. positive Lyapunov exponents have been computed for both the discrete logistic model and the discrete Lorenz model (Section 3.4.3), Werndl (2009b, p. 211) points out that this precise form of divergence is not necessarily associated with apparent stochasticity (or other features of chaotic models).

3.4.5   Strange Attractors

In Section 3.3.2, attractors were introduced as a natural means of restricting the region of phase space considered, that is as a condition requiring that a given set of criteria for chaos only needs to be fulfilled in a certain region of phase space. We recall: an attractor is a region of phases space to which all nearby trajectories or points of a model eventually tend. For the behaviour of two of the models used as case studies in this book – namely, the iterated logistic model for values of $r > 4$ (Section 2.3.3); and the iterated Lorenz model (Section 2.4.2) – the existence of an attractor has been formally established. In the discrete Lorenz model (Section 2.4.1), a particular butterfly-shaped phase space region, which seems to be attracting, can easily be visually distinguished. In Section 4.5, I will discuss how investigative work with two related horizontal models also provides formal evidence towards this region being a true attractor.

If attractors are solely interpreted as phase space restrictions, then they are not a part of the requirements of a chaos definition but should rather be viewed as specifying of the domain of a given definition. Numerous examples of this use of the concept can be found in the chaos literature (e.g. Werndl, 2009b, p. 207, Alligood et al., 1997, p. 240).

However, other authors appear to use the existence of an attractor as an explicit requirement in a chaos definition. For example, Hastings et al. (1993, pp. 6) write:

> Another characteristic of chaotic behaviour is the existence of a strange attractor to which all sufficiently nearby solutions eventually converge, given sufficient time.

Other authors also treat the existence of an attractor as an explicit requirement for a diagnosis of chaos (e.g. Smith, 1998, p. 13; and, somewhat less explicitly, Ruelle, 1991, pp. 9–10; Ott et al., 1994, pp. 9–11). As discussed in Section 3.3.3, the requirement that an attractor exists can be viewed as a weak embodiment of the criterion of periodicity. In addition, all authors discussed in this section also require the models they investigate to be deterministic.

The mere existence of an attractor does not have sufficient traction as a distinguishing criterion for chaos since many deterministic, periodic systems also possess attractors. Therefore, the embodiment is usually either (i) combined with a further requirement of periodicity, that is a strengthening of the embodiment and/or (ii) another criterion for chaos. Attractors which fulfil such additional requirements are often given the label "strange". However, due to the fact that this strengthening of the embodiment or the combination of additionally required criteria can vary across different definitions, the attribute "strange" can assume various meanings. In fact, definitions of chaos through the existence of a strange attractor are so variable, that their extensions differ significantly from each other. In contrast to my previous analyses of chaos definitions in this chapter (Section 3.4.1– 3.4.4), during which it was possible to outline a core set of models fulfilling a given definition, I will therefore forego a discussion of the extension of the definition of chaos in terms of the existence of strange attractors, and instead outline the most prevalent versions of this definition as separate concepts.

Requirement of Fractal Geometry

Alternative (i), that is a strengthening of the embodiment by requiring further periodicity, usually takes the form of requiring fractal geometry for the attractor, which can be determined by showing either self-similarity or the existence of a fractal dimension (Ott et al., 1994; Smith, 1998, e.g.). Self-similarity, roughly, describes a similarity of the behaviour of a model on several spatial (or otherwise defined) scales. This criterion of self-similarity is fulfilled by the iterated logistic model (Section 2.3.3, Fig. 2.3), whose attractor is the self-similar Cantor set. A similar fractal set of fixed points has also been shown to exist in the iterated Lorenz model (Section 2.4.2). For numerical models, for example, the discrete Lorenz model (Section 2.4.1), a fractal dimension can be assigned by comparing how much the attracting region extends in each dimension of phase space. The butterfly-shaped Lorenz attractor is thereby usually assigned a fractal dimension slightly exceeding 2.0 (Alligood et al., 1997, p. 366), which corresponds to the visual impression that the attracting region is almost two-dimensional. Similar attractors have been found both in a number of other numerically integrated models as well as in experimental data sets (e.g. for a collection of studies, Ott et al., 1994, chapter 7).

Chaos definitions that are based solely on the requirement of the existence of an attractor with fractal dimensionality have been criticized as not capturing the notion of irregularity and unpredictability, which in most definition is crucially associated with idea of chaos (e.g. Lorenz, 1993, p. 137). Definitions of chaos which rely on the requirement of the existence of a fractal attractor by itself are therefore rare.

Requirement of Other Criteria

In most cases, the requirement of the existence of an attractor is paired with a second phenomenological criterion for chaos, which puts demands on the behaviour of the model on the attractor (e.g. Ruelle, 1991; Hastings et al., 1993; Smith, 1998). In this context, the two criteria most often required are SDIC (Section 3.3.5), often in the embodiment of a requirement of positive Lyapunov coefficients, and aperiodicity

(Section 3.3.4), often in the embodiment of apparent stochasticity. While it has been argued that definitions that require a strange attractor in combination with other criteria for chaos are essentially versions of chaos definitions based on these criteria, in combination with a restriction of the relevant domain (e.g. Werndl, 2009b, p. 207), it is often apparent from the discourse in which such combinations are used that they are meant to capture a particular pre-formal notion of chaos: namely, one which views chaotic behaviour as a combination of elements of irregularity and regularity. An example of such a conception of chaos can be found in Smith (1998, pp. 13), who describes chaos as requiring a "combination of large-scale order with small-scale disorder". The persuasiveness of arguments for explicitly including the requirement of the existence of an attractor into definitions of chaos therefore depends on how valuable this pre-formal idea about chaos is judged to be.

The extensions of definitions of chaos based on the requirement of the existence of a strange attractor overlap with the extensions of other chaos definitions in various ways, depending on the additional phenomenological criteria they require. Many two-dimensional models that have been judged chaotic according to other definitions possess fractal attractors: for example, the two iterated models used as case studies in this book, Sections 2.3.3 and 2.4.2; and the class of models similar to Smale's horseshoe map. However, a significant drawback of such definitions of chaos is the exclusion of a number of paradigmatically chaotic models: for example, the discrete logistic model (Section 2.3.2); the iterated logistic model for $r = 4$ and similar one-dimensional maps, all of which display chaotic behaviour according to other definitions but do not possess attractors (e.g. Werndl, 2009a, p. 208; Smith, 1998, p. 168). In much of chaos literature, attractors therefore seem to be viewed as a common feature in chaotic models, and as a natural means of restricting the domain of applicability of a given chaos definition, but not as an embodiment of a criterion for chaos.

3.5 CONCLUSION

As outlined in Section 3.1, this chapter has two main methodological purposes: (i) the introduction of a framework for the rational reconstruction of chaos definitions and (ii) a demonstration of the merits of this framework by using it to analyse the structures and uses of different chaos definitions. During the analytic task (ii), I obtained a number of results that will be important in my discussion of the evaluation of chaotic model in Chapter 4. With respect to task (i), I presented an analytic framework that is based on the assumption that chaos definitions can be decomposed into requirements of different combinations of five core criteria for chaos (Section 3.3): the dynamical criterion of determinism (Section 3.3.1); and the four phenomenological criteria of transitivity (Section 3.2.2), periodicity (Section 3.3.3), aperiodicity (Section 3.3.4) and SDIC (Section 3.3.5). Each of these criteria can be conceptualized as a similarity category, which contains a number of more concrete, formal embodiments of the concept. Different embodiments of a criterion will thereby be geared towards the technical specifications of different types of models. In this framework, chaos definitions can vary in two ways: (i) in the combination of general criteria they require; and (ii) in the choice of specific embodiments of these criteria.

The concept of "chaos" as used by practitioners and philosophers in chaos theory appears to be solidly grounded in the five criteria but can also be given different realizations in different definitions. The co-existence of different chaos definitions therefore does not appear to be a consequence of fundamental disagreements about how the notion should be conceptualized or about the set of models on which the label "chaotic" should be bestowed. Instead, it appears to be a consequence of the complex interplay of different models in chaos theory, which has already been outlined in Chapter 2 and will be discussed further in Chapter 4. Different chaos definitions are usually particularly applicable to a specific class of models and they appear to highlight precisely those properties of these models that are particularly important to the models' investigative uses.

In Section 3.3, I described the core notion behind each criterion and its most prevalent embodiments. It thereby became apparent that, for most criteria, there are a large number of different embodiments: in particular, the criterion of aperiodicity, which is in itself a contested concept, has been cast into many different embodiments, some of which are shared with other criteria. The relationships between different embodiments of different criteria can be complex. For example, some combinations of embodiments of other criteria for chaos imply embodiments of SDIC.

One important conceptual result of the discussion in Section 3.3 is the identification of determinism as a dynamical (pre)-criterion for chaos. Interpreting determinism as a criterion for the diagnosis of chaos rather than a necessary or sufficient condition for the occurrence of chaotic behaviour appears to be closer to the actual usage of the concept by practitioners and will also be shown to allow for a clearer exposition of the difficulties encountered during the determination of the existence of chaos in natural systems (Section 4.3.2).

I also argued that the local and global forms of unpredictability associated with chaos should be viewed as embodiments, or consequences, of the criteria of aperiodicity (Section 3.3.4) and SDIC (Section 3.3.5), respectively. My analysis therefore provides additional support for the claim that unpredictability is an associated, but not a defining, feature of chaos (e. g. Werndl, 2009b; Batterman, 1993).

Furthermore, I was able to use the double-decomposition into criteria and embodiments to comparatively analyse five prevalent chaos definitions: Devaney chaos (Section 3.4.1); (the definition of chaos as) mixing (Section 3.4.2); (the definition of chaos in terms of) positive Lyapunov exponents (Section 3.4.3); stochastic chaos (Section 3.4.4) and (the definition of chaos in terms of) strange attractors (Section 3.4.5). Each of these definitions requires different combinations of criteria (Table 3.2) in different embodiments. The similarities and differences between these combination of criteria have been shown to provide a good heuristic guide to estimating overlaps between the extensions of different definitions. It has also become apparent that there is considerable overlap between the extensions of different chaos definitions.

My analysis of the relationships of different criteria (Section 3.3) and different definitions of chaos (Section 3.4) shows that the differences in the domain of applicability, and the technically complex relationships between different embodiments of different chaos criteria, greatly mitigate the demarcating powers of any one definition of chaos. Scenarios in which two different chaos definitions are directly applicable to one particular model, and the model fulfils the requirement of on definition but not of the other, are therefore rare, although they can be constructed (e.g. Smith, 1998, Chapter 10). The hypothesis that the coexistence of different chaos definitions is practically useful and seldom leads to acute conflicts of classification is also borne out by the quietism with which most practitioners acknowledge their existence (e.g. Berger, 2001, p. 40, Hirsch et al., 2004, chapter 15).

The view that "chaos" is a malleable concept both in its precise conceptual requirements and in its technical specifications, which arises from my analysis in this chapter, appears to be compatible with Kellert's (1993) description of chaos theory as a tool box, which offers scientists the means to investigate certain, previously ignored, properties of models and natural systems. Under this interpretation, the different chaos definitions can be viewed as investigative tools, which highlight the properties of models that will be most important to their investigative purposes. In Chapter 4, the investigative use of horizontal models in chaos theory and their relationships with vertical models in their lineage will be analysed in detail and the function of particular chaos definitions as a means to highlight particular properties relevant to such investigations will be further illustrated.

As outlined in Section 3.1, I do not claim that the framework presented here is the only conceivable one for the analysis of chaos definitions. However, I am confident that the framework is a useful one and hope to have demonstrated some of its merits by deriving a number of results pertaining to unresolved questions on the specific nature of chaos definitions and by offering a novel view of the use of definitions in chaos theory.

CHAPTER 4

Evaluation of Chaotic Models

Abstract I will introduce an analytic framework that conceptualizes the evaluation of vertical chaotic models as consisting of three steps: the determination of the chaotic conditional to be transferred from a model to a target system; the determination of the existence of chaos in the target system; and the evaluation of model faithfulness. Each step will be discussed in detail. I will also discuss the evaluation and investigative role of horizontal chaotic models.

Keywords chaos · model faithfulness · conditions for chaos · models in science

4.1 INTRODUCTION

In this chapter, I will discuss the evaluation of chaotic models. I will thereby return to Suarez's (2013) analytic framework based on the transference of conditionals (Section 2.2.1). In this section, I will adapt this framework for the specific evaluation of chaotic models.

4.1.1 A Three-Step Framework for the Evaluation of Vertical Chaotic Models

In Section 2.2.1, I gave a general account of the evaluation of vertical models. I introduced a framework for the analysis of such evaluations, which is based on the notion that the evaluation of vertical models consists

L.C. Zuchowski, *A Philosophical Analysis of Chaos Theory*, New Directions in the Philosophy of Science, DOI 10.1007/978-3-319-54663-6_4

in a comparison of the model system with the target system and, in particular, aims to determine whether conditional relationships C → B established in the former can be transferred to the latter (Suarez, 2013). In this context, I also introduced the notion of model faithfulness (Definition 1), which asserts that a behaviour B is faithfully modelled if no fictional aspects of the model system are part of the posited sufficient conditions C for this behaviour B.

In the specific case of the evaluation of chaotic models, I maintain that his evaluation process can be further broken down into three separate conceptual steps:

i. First, the conditional C → B to be transferred from the model system must be determined, that is practitioners engaged in evaluating a model need to decide which chaotic behaviour B exists in the model and which conditions C are sufficient for this behaviour. They thereby need to provide suitable evidence for the claim that, if the conditions C are met, then the behaviour B will occur in the model.

ii. Second, practitioners will only consider the transference of a conditional C → B if the chaotic behaviour B has been shown to exists in both the model and the system (Section 2.2.1). They therefore need to determine that the target system also displays the chaotic behaviour B.

iii. Third, the faithfulness of the model with respect to the behaviour B should be determined, that is practitioners should decide whether the sufficient conditions C contain any fictional parts.

As with the twofold decompositional framework for the analysis of chaos definitions in Chapter 3, this analytic framework is meant as a means of rational reconstruction. I intend to demonstrate its merits as an analytic tool in the sections to come. However, I do not claim that each step is fully articulated in each practical evaluation of a chaotic model, that is, the framework is descriptively precise, neither do I wish to propose this three-step procedure as normatively desirable. However, following Suarez (2013), I think that this inferential framework captures the most important conceptual elements of the evaluation of vertical chaotic models as documented in the literature. In Section 4.3, I will also show that its adoption allows a clear identification and description of particular problems at each stage (i)–(iii) and hence enables the philosopher to make some low-key normative recommendations for the resolution of the issues identified.

A first comparison between this framework and the approaches to the evaluation of chaotic models taken in the philosophical (e.g. Koperski, 1998; Smith, 1998; Kellert, 1993) and scientific (e.g. Guckenheimer et al., 1977; Pool, 1989) literature shows that the existing literature focuses almost exclusively on step (ii) in the evaluation process, that is the determination of whether seemingly chaotic behaviour in a natural system is really chaotic or should be given a different label (e.g. that of "noise"). The framework underlying these studies seems to be a conceptualization of model evaluation as a bit-by-bit comparison: it is assumed that the evaluation of a vertical model consists in a straightforward comparison of the model's output with the behaviour of the target system (in a suitable representation).

I hope to add to the existing literature by using the inferential framework to identify and localize the particular issues that arise during each step of the evaluation of chaotic models. Furthermore, the framework will enable me to analyse in detail the investigative role of horizontal models in chaos theory (Bokulich, 2003, Section 2.2.2): I will show that these models are used as investigative tools in the determination of conditionals $C \rightarrow B$ to be transferred from related vertical models in their lineages (Sections 4.3.1 and 4.4). As described in Section 3.1, the framework also enforces a clear distinction between the sufficient conditions C for the occurrence of a behaviour B and the criteria for the diagnosis of this behaviour. Since both the conditions for and the definitions of chaos are still contested subjects, I hope to demonstrate that this clear distinction leads to a clearer exposition of both topics.

4.1.2 Outline of the Chapter's Content

In Section 4.2, the three-step framework will be used to analyse the evaluation of the two chaotic logistic models (Sections 2.3.2 and 2.3.3). Analogous to their function in previous chapters, these case studies will be used to motivate and illustrate the general discussion of the evaluation of chaotic models in later sections of this chapter.

In Section 4.3, each of the three conceptual steps in the evaluation of vertical chaotic models will be discussed in detail. In Section 4.3.1, the determination of the conditional $C \rightarrow B$ to be transferred from the model will be analysed. I will argue that a large number of different conditionals are used by practitioners. This variety in the conditionals $C \rightarrow B$ used in the evaluation of chaotic models is a consequence of the large number of coexisting chaos definitions (Chapter 3), which allows for the specification of different chaotic behaviours B. However, I will then argue that there are generally only two different types of sufficient conditions C assigned to these behaviours B: forms of non-linearity and discretization (type 1); and forms of non-linearity and iteration (type 2). The conditionals to be transferred from chaotic model systems to their target systems can therefore be divided into two types, each corresponding to a different set of sufficient conditions C.

It will become apparent that the determination of the conditional $C \rightarrow B$, that is the determination of whether if C is true in the model, B will be true as well, often involves considerable investigative work with related horizontal models.

I will also briefly discuss the determination of such conditionals under the interpretation of chaos as defined by the existence of global unpredictability (which I rejected for independent reasons in Section 3.3.5) and show that it requires the positing of the existence of an error source as a sufficient condition for chaos. In Sections 4.3.2 and 4.3.3, I will then show that this implies that chaos (under this definition) can never be modelled faithfully and that these accounts appear to be incompatible with an interpretation of chaos as a natural property.

Furthermore, I will use the identification of the different conditionals used in chaos theory to highlight a particular feature of the evaluation of numerical chaotic models: I will show that while these models are discrete models (e.g. Section 2.4.1), the conditionals $C \rightarrow B$ practitioners try to establish are of type 1, that is, they do not have discreteness as a sufficient condition C for a chaotic behaviour B. In Section 4.5, the fact that such determinations are often difficult and require considerable investigative work will be illustrated in a case study of the evaluation of the discrete Lorenz model.

In Section 4.3.2, I will discuss the determination of the existence of chaos in the relevant natural target systems, that is, the completion of the second step of the evaluation process. It has generally been recognized that the detection of chaos in nature poses unique problems (e.g. Smith, 1998; Koperski, 1998; Lorenz, 1993; Pool, 1989). The majority of these problems arise in the context of determining the existence of stochastic chaos (Section 3.4.4). Since stochastic chaos is phenomenologically indistinguishable from the behaviour of a truly stochastic system (i.e. a Bernoulli process) by definition, the observation of stochastic behaviour in a target system is not sufficient to determine that the system is really stochastically chaotic (rather than "noisy" or otherwise probabilistic). Based on the results of my analysis of chaos definitions in Chapter 3, I will reinterpret this problem as a difficulty of proving the fulfilment of the dynamical (pre-)criterion of determinism (Section 3.3.1). Technically, showing the fulfilment of the criterion of determinism from a given set of empirical data involves extracting information about the dynamical properties of a system from data about its phenomenology. Since the behaviour of the system during the chaotic phase cannot be used towards this purpose, attempts to solve this problem usually utilize additional information about the non-chaotic phases of the model and the target system.

In Section 4.3.3, I will discuss the determination of the faithfulness of vertical chaotic models, that is the completion of the third step in the evaluation process. I will show that, for most models with type 1 and type 2 conditionals, this is a relatively straightforward, albeit sometimes neglected step in the evaluation process. However, a consideration of the model faithfulness of chaotic numerical models, that is discrete models of a continuous set of governing equations, will highlight the fact that only type 1 conditionals, that is conditionals that posit a form of iteration on a set of continuous trajectories, rather than a form of discretization, as a sufficient condition C for a chaotic behaviour B, can be modelled faithfully by numerical models. Accordingly, much of the modelling activity in chaos theory can be interpreted as part of an effort to determine such conditionals for numerical models, a task that is made imperative by the fact (as explained above) that discreteness is a known fictional feature of these modes and should hence not feature in the antecedents of chaotic conditionals to be transferred.

In Section 4.4, I will discuss the evaluation of horizontal chaotic models. I will show that the first step of the three-step evaluation process described above, that is the assignment of sufficient conditions to a particular kind of chaos observed in the model, is often part of the evaluation of horizontal models as well. However, the conditionals established in this way are not meant to be transferred to natural systems; rather, their determination should be viewed as part of an investigation of the models' properties. The main focus of this section will be on a discussion of the investigative role of horizontal models during the first step of the evaluation process of related vertical models: namely, their use in determining the appropriate conditional C → B to be transferred from those related models to their target systems. The section therefore serves both to illustrate the interplay between horizontal and vertical chaotic models as well as to highlight the difficulties that arise when inferring properties of the latter from investigative work with the former. Since the notion of horizontal modelling has not been used in the context of chaos theory yet, my analysis in this section is novel. It also provides additional evidence for the importance of horizontal models as argued for by Bokulich (2003).

In Section 4.5, I will apply results from the discussions presented in Chapters 2, 3 and 4 to the case study of the evaluation of the Lorenz lineage (Section 2.4). In particular, I will use the analytic tools developed in these chapters to provide an in-depth philosophical analysis of Smale's 14th problem, that is the question of whether the discrete Lorenz model (Section 2.4.1) and the iterated Lorenz model (Section 2.4.2) are chaotic in the same way (Smale, 1998). I will argue that attempts to solve this question involve both the construction of a third horizontal model, the rigorous Lorenz model as well as the use of this model to provide evidence for the fact that a type 1 conditional should hold true in the model, that is an iterative dynamical process rather than the discrete nature of the model should be viewed as a sufficient condition for the appearance of chaos in the model.

4.2   Evaluation of the Lineage of Logistic Models

In this section, I will describe the evaluation of the two chaotic logistic models (Section 2.3). As in previous chapters, these case studies will have the double function both of illustrating the more general discussions in (Section 4.3 and 4.4 and as well as of introducing some of the key analytic concepts to be used in this chapter.

In Section 4.2.1, the evaluation of the vertically constructed, discrete logistic model (Section 2.3.2) will be discussed. My analysis will use the three-step conceptualization of model evaluation developed in Fig. 4.1. It will be shown that an indirect approach was taken to the evaluation of this model, leading to the conclusion that no direct evaluation is necessary since the model does not actually predict the occurrence of chaos in its target system. In Section 4.2.2, I will discuss the evaluation of the horizontally constructed, iterated logistic model (Section 2.3.3). I will then show that this model has been used to establish the existence of a particular conditional C → B in the discrete logistic model.

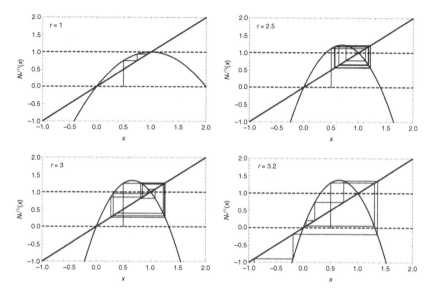

Fig. 4.1 Construction of the discrete logistic model N_x from the first iteration of the (shifted) iterated logistic model $N_*(x)$. The cobweb diagram is obtained by projecting a value between the graph $N_*(x)$ and the identity function $N(x) = x$. The construction uses an initial value $N_0 = 0.5$. The unit interval is indicated by dashed lines

4.2.1   Evaluation of the Discrete Logistic Model

While May (1974) and May and Osler (1976) provide some evaluation of the discrete logistic model, neither of these studies contains any comparison of the model to laboratory or natural systems of non-overlapping populations (Section 2.3.2). The first detailed evaluation against such systems was provided by Hassell et al. (1976). This case study will therefore be based on an analysis of all three of these papers.

Determining the Conditional to Be Transferred
Using the three-step framework introduced in Section 4.1, I will first discuss the determination of the conditional C → B to be transferred from the discrete logistic model. As described in Section 3.2.2, the type of chaos diagnosed by May (1976) and May and Osler (1976) is stochastic chaos (Section 3.4.4), which is based on the criteria of

aperiodicity (Section 3.3.4) and determinism (Section 3.3.1), and implies the criterion of SDIC (Section 3.3.5). The behaviour B in the conditional to be evaluated is therefore stochastic chaos (Table 3.2).

The antecedent C in the conditional C → B must be more carefully reconstructed. Hassell et al. (1976, p. 471) explicitly name one sufficient condition for stochastic chaos in the discrete logistic model:

> It is now appreciated that density dependence alone cannot only lead to stable equilibria, but also can lead to patterns of stable cycle oscillations (with period and amplitude determined by the intrinsic biological parameters), and even to irregular and apparent 'chaotic' population fluctuations.

Therefore, one part of the antecedent C in the conditional is "density dependence", that is the fact that model (2.4) quadratically depends on the population number itself.

While neither of the authors explicitly discusses further conditions for chaos (May and Osler, 1976, pp. 574–577) review results from an investigation with the iterated logistic model by Li and Yorke (1975), which provides evidence that the apparently stochastic behaviour of the discrete logistic model can be seen as the result of an alternative construction process that is based on following the fate of single points on the iterated logistic model. In Section 4.2.2, Li and Yorke's (1975) study, and a similar investigation by Guckenheimer et al. (1977), will be discussed in detail. In this section, it will be shown that these investigations provide persuasive evidence to view a form of discretization as part of the sufficient conditions for chaos in the discrete logistic model. However, during the early evaluations of the discrete logistic model, the role of discretization remained unresolved: despite their presentation of the material from Li and Yorke (1975), the authors surmise that some continuous models might show stochastic chaos as well (e.g. May and Osler, 1976, p. 574).

May (1974), May and Osler (1976) and Hassell et al. (1976) therefore determine the following conditional for transference to a target system:

Conditional 1. Logistic (quadratic) form ∧ (Form of Discretization) → Stochastic chaos.

While these evaluations of the discrete logistic model left the role of discreteness open, and this condition is therefore shown in brackets in Conditional 1, we will see that there exists investigative work with the iterated model that can be interpreted to establish it more firmly as a sufficient condition (Section 4.2.2). As discussed in Section 2.2.1, determination of the conditional implies that it has either been shown or can reasonably be assume that if the sufficient conditions C are present in the model, so will be the behaviour B.

Determining the Existence of Chaos

The second step in the evaluation pro-cess (as conceptualized by my analytic framework) is the determination of the existence of chaos in the target system. Hassell et al. (1976) perform this step in the evaluation of the discrete logistic model. They compare the behaviour of the model to a number of data sets collected from laboratory and natural insect popula-tions with discrete generational dynamics. The strategy they use to deter-mine whether these data sets are chaotic is an indirect one, which I will outline below.

First, a general form of the discrete logistic model is established (Hassell et al., 1976, pp. 472–474):

$$N_{x+1} = \lambda N_x (1 - a N_x)^{\beta} \qquad (4.1)$$

where a and β are free parameters, which can vary for different insect populations. This general model is analysed and parameter regimes of λ and β in which the model displays (visual) stochastic chaos are identified (Hassell et al., 1976, Fig. 2).

Second, the model (4.1) is fitted onto a number of experimental and laboratory data sets (obtained from studies published between 1961 and 1974). Twenty-four different data sets are fitted and, for each of these, the parameters λ and β, are estimated by a best fit algorithm.

Third, these estimated parameters pairs are checked against the model's regime diagram to determine whether they fall into the chaotic regime. If this were the case, then the authors would report this as evidence that the target system could be chaotic and that Conditional 1 should be consid-ered for transference to the natural system under investigation.

While the study contains one image displaying actual population fluc-tuations of a laboratory culture of beetles (Hassell et al., 1976, Fig. 3), no attempts are made to directly determine whether the recorded time-series are stochastically chaotic, that is no attempts are made to show that they fulfil any embodiments of the criteria for stochastic chaos (Section 3.4.4). Instead, the authors only determine whether these systems possess λ and β parameters that fall into the chaotic regime of the model.

This indirect approach is retrospectively justified by the fact that Hassell et al. (1976) come to a negative conclusion: none of the fitted populations possesses λ or β values that would clearly place it into the model's chaotic regime; only two populations fall within the high periodicity regime that precedes the chaotic phase in the discrete logistic model (Section 2.3.2). The vast majority of observed populations appear to be best modelled by versions of (4.1) that only show stable oscillatory or monotonic behaviour. Hassell et al. (1976, p. 483) therefore conclude:

> [W]e conclude that some of the types of dynamical behaviour that are possible in theory, in fact, rarely occur in real, single-species populations.

The conclusion that the model does not actually predict stochastic chaos for these target systems implies that the question of whether Conditional 1 transfers to these systems is mote. It also allows Hassell et al. (1976) to forgo directly determining the existence of stochastic chaos in the data collected from the target systems.

Determining Model Faithfulness
Since the second step of the evaluation process concludes with a negative result, and the transference of Conditional 1 is therefore no longer being considered, Hassell et al. (1976) do not perform an explicit evaluation of model faithfulness (Definition 1). The third step of the evaluation process is therefore not completed in this case study. Instead, the author's discussion section focuses on outlining aspects of the target systems that might have made these systems less likely to display chaos. I will discuss the two factors that Hassell et al. (1976, pp. 483–484) judge most important in this regard below.

First, they discuss the fact that the model applies to single-species populations only, which is a good description of laboratory target systems but will be an idealization for natural systems. A multi-species model would therefore be a better fit for this latter class of target systems and such systems are hypothesized to be more likely to be located in the appropriate models' chaotic parameter regimes.

Second, the authors point out that the model only applies "to the restricted range of organisms with completely discrete generations" (p. 483). The insect populations they have chosen as target systems fulfil these conditions. However, Hassell et al. (1976) maintain that a class of models targeted towards systems with overlapping generations might have more potential to fall within a chaotic regime than the discrete models discussed in this study. In particular, they name a class of models based on time-continuous differential equations with time-delays as probable candidates for displaying chaotic behaviour "if the non-linear dynamics are complicated enough" (p.484)

Time-delay models are described as "time-continuous" by both Hassell et al.(1976, p. 484) and May and Osler (1976, p. 574). However, the use of "continuous" can be misleading in this context: it "hides" the fact that, while the governing equations of the model are time-continuous, the model itself will usually be discrete. Time-delay models with complicated non-linear dynamics can usually not be solved analytically (e.g. Cushing, 1977). Since the construction of numerical solutions involves discretization (e.g. Sections 2.3.2 and 2.4.1), the actual models in this scenario would then be discrete rather than continuous.

However, in contrast to the case of the discrete logistic model, discreteness is a fictional part of a time-delay model. Accordingly, chaotic behaviour B that is part of a conditional C → B similar to Conditional 1, that is a conditional with discreteness as part of the sufficient conditions C, would not be modelled faithfully. Therefore, Hassell's et al. (1976, p. 484) and May' and Osler's (1976, p. 574) emphasis on the time-continuous nature of time-delay models seems to reflect the necessity to avoid positing discreteness as a necessary condition for chaos in these models. I maintain that this necessity to establish that discreteness is not a necessary condition for chaos, that is there are other sufficient conditions for the occurrence of chaotic behaviour, is a general characteristic of the evaluation of numerical models in chaos theory. It will be discussed in more detail in Section 4.3.1. Since research on chaos in time-delay models of insect populations has not yielded the hoped for clear evidence for the existence of chaos in these populations (e.g. Hastings et al., 1993), the question of which conditionals transfer from these models has not been subject to further investigations. However, in Section 4.5, I will present a case study, namely the evaluation of the discrete Lorenz model, in which related horizontal models have been used extensively to rule out the existence of conditionals similar to Conditional 1 in a numerical model.

While an explicit evaluation of model faithfulness has not been provided in the literature, I will argue in Section 4.3.3 that stochastic chaos (and related kinds of chaos) are modelled faithfully by the discrete logistic model. Since this discussion relies on a version of Conditional 1 in which discreteness has been affirmed as a sufficient condition of chaos through investigative work with the iterated logistic model (Section 4.2.2, Conditional 3), this further discussion of the model faithfulness of the discrete logistic model has been moved to a later part of the chapter.

Further Evaluation of Chaotic Models in Population Dynamics
Hassell's et al. (1976) negative result on the of potential of insect populations to develop stochastic chaos appears to have been representative for these populations and has been confirmed by a number of similar studies. Since the late 1970s, there have been numerous studies on the existence of stochastic chaos, positive Lyapunov coefficients or other embodiments of SDIC in more complex population models and systems. The most prominent class of these models represents discrete population dynamics on a spatial grid (e.g. Kot and Schaffer, 1986; May, 1994). However, the questions of whether (i) the target systems of these models are located in chaotic parameter regimes; (ii) chaos of some definition is really present in these target systems and (iii) chaos is modelled faithfully by these models remain controversial (e.g., for review, Pool, 1989; Hastings et al., 1993).

Given the indirect nature of Hassell's et al. (1976) approach, one might think that the evaluation of the discrete logistic model is not indicative of the evaluation of other vertical chaotic models. To some extent, this is true: in Section 4.3.2, I will show that the data-fitting methodology used by Hassell et al. (1976) has been replaced by other diagnostic techniques, most prominently that of phase-space reconstruction. However, the case study is a good illustration of two general features of the evaluation of vertical chaotic models: (i) it demonstrates the importance placed by practitioners (and philosophers) on the second step in the evaluation process, that is on the determination of the existence of chaos in the target system; and (ii) it highlights the investigative use of related horizontal models during the first step of the evaluation process, that is during the determination of the conditional C → B to be considered for transference. In the next section, the latter aspect (ii) of the evaluation of models in the logistic lineage will be discussed in more detail. In Sections 4.3.2 and 4.3.1, respectively, I will discuss both features (i) and (ii) as general characteristic of the evaluation of chaotic models.

4.2.2 Evaluation of the Iterated Logistic Model

As discussed in Section 2.2.2, horizontal models are not evaluated against target systems. Instead, they are used to explore the mathematical properties of other models in their lineages. The evaluation of the iterated logistic model, whose horizontal construction I described in Section 2.3.3, is a good example of such an investigative use of a model. In particular, I will show that: (i) a chaotic conditional C → B is established during the evaluation of the iterated logistic model, but that this conditional is not considered for transference to a natural system; and (ii) that the iterated logistic model is instrumental in confirming that discretization should be included as a sufficient condition in Conditional 1, that is it is instrumental in determining the conditional to be transferred from the related – vertical – discrete logistic model. It will also become apparent during this discussion that the definition of Devaney chaos highlights precisely those features of the iterated logistic model that are important during its investigative use. Accordingly, the case study also illustrates the epistemic usefulness of the co-existence of different chaos definitions (Sections 3.4 and 3.5).

Determination of the Chaotic Conditional
As we have seen in Section 3.2.3, the iterated logistic model is Devaney chaotic (Section 3.4.1). The conditions under which this and similar models are expected to show this behaviour have been investigated, that is there have been attempts to determine a conditional C → B with Devaney chaos as the behaviour B. In contrast to the evaluation of the discrete logistic model (Section 4.2.1), the determination of this conditional appears to be solely a matter of mathematical exploration: it aims at establishing the sufficient conditions for chaos in the model but not at transferring this conditional to a natural target system.

The sufficient conditions C that are usually assigned to the appearance of Devaney chaos in the iterated logistic model are already apparent in the discussion of the behaviour of the model in Section 2.3.3: for cases with $r > 4$, the behaviour can be viewed as a result of the iterative "removing the middle" procedure; in the case of $r = 4$, it seems to be more appropriate to consider the iterative "folding" of the function into the unit interval a condition of the chaotic behaviour (Fig. 2.2). As stressed in Section 2.2.1, acceptance of the conditional only implies the belief that, if such an iterative process is present in a model sufficiently similar in other aspects to the iterated logistic model, then Devaney chaos will be present as well.

The belief that an iterative procedure, conceptualized either as the self-application of the function $N(x)$ or the construction of a Cantor set or a topological "folding" of the function, is a sufficient condition for chaos in the iterated logistic model has been expressed by several authors (e.g. May and Osler, 1976, p. 584; Devaney, 1989, pp. 139–140; Smith, 1998, p. 46). The presence of a form of iteration will therefore be one of the sufficient conditions C in the conditional C → B that holds true in the iterated logistic model.

Furthermore, there is a general consensus that such iterative procedures, like folding or self-application, will only produce chaotic behaviour if the iterated sequence (or function) is non-linear (e.g. Hirsch et al., 2004, pp. 140–141; Smith, 1998, Chapter 1). Therefore, the second sufficient condition C for Devaney chaos in the iterated logistic model is the non-linear nature of the model's dynamics. This leads to the following conditional:

Conditional 2. $\dfrac{\text{Non} - \text{linear folding sequence}}{\text{Non} - \text{linear (quadratic) function}} \dots \wedge \text{Iteration} \rightarrow$ Devaney chaos.

Conditionals which posit a form of non-linearity and an iterative procedure as sufficient conditions for a particular kind of chaos will be called type 1 conditionals. In Section 4.3.1, I will discuss this type of conditional in more detail.

In the first instance, the conditional is model specific, that is it only states that the combination of these two ingredients in the iterated logistic model's dynamics are seen as sufficient conditions for the appearance of Devaney chaos in this model. Conditional 2 should not be taken to mean that all models which are iteratively constructed from non-linear equations are also Devaney chaotic. However, the determination of such conditionals in horizontal models is used in the investigation of other similar models by providing guidance towards the identification of the sufficient conditions for chaos in these models. For example, in the case of the iterated logistic model, Conditional 2 is generalized to all models based on quadratic equations (e.g. Hirsch et al., 2004, pp. 337–342). However, it is apparent from such discussions that both the existence of chaos as well as the likely generating factors need to be established anew for each model (e.g. Hirsch et al., 2004, p. 356; Devaney, 1989, pp. 51–53). This illustrates the work intensive nature of modelling in chaos theory, which will also become apparent during the general discussion of the evaluation of chaotic models below (Section 4.3.1 and 4.3.3)

Investigative Use of the Iterated Logistic Model

The two seminal works in the evaluation of the iterated logistic model and of its relationship with the discrete logistic model are Guckenheimer et al. (1977) and Li and Yorke (1975). Both papers provide additional evidence for the inclusion of discretization as an antecedent in Conditional 1 by highlighting alternative ways of constructing the discrete logistic model from a shifted version of the iterated logistic model through two particular means of discretization. In other investigations, further evidence for the inclusion of discretization as a sufficient condition in Conditional 1 has also been provided through the use of other models (e.g. Tsonis, 1992, pp. 132– 136). In this section, I will only discuss the investigative use of iterated logistic model.

Li and Yorke (1975) formally establish that – as a consequence of Sarkovskii's theorem (Theorem 1) – iterated maps that feature a periodic point of prime period 3 will also possess periodic points of all other prime periods. Since the iterated logistic model can be shown to have periodic points of prime period 3, the large numbers of periodic points displayed in Fig. 2.3 can be viewed as a consequence of Sarkovskii's theorem. Li and Yorke (1975, p. 987) then show that such maps also have an uncountable number of points that are not periodic and that do not approach periodic points even for large numbers of iterations. This is also apparent in Fig. 2.3, where, for example, the points located between fixed points are (eventually) mapped to minus infinity. It is easy to see that this part of the investigation crucially depends on the dense periodicity and transitivity highlighted in the definition of Devaney chaos (Definition 5, Section 3.4.1).

Li and Yorke (1975, p. 985) now link these features of the iterated logistic model to the behaviour of the discrete model:

$$N_{x+1} = rN_x(1 - N_x). \tag{4.2}$$

This discrete model is not identical to the discrete logistic model (2.4), which is obtained through direct discretization of the logistic equation and which has been investigated by May (1974).

However, the results from Li and Yorke (1975) and Guckenheimer et al. (1977), who also use the discrete model (4.2), directly transfer qualitatively to the discrete logistic model and will transfer quantitatively if the iterated model used is the following:

$$N_*^{(0)}(x) = x,$$

$$N_*^{(1)}(x) = N_*^{(0)}(x)\left(1 + r\left(1 - N_*^{(0)}(x)\right)\right)$$

$$\ldots$$

$$N_*^{(n+1)}(x) = N_*^{(n)}(x)\left(1 + r\left(1 - N_*^{(n)}(x)\right)\right) \qquad (4.3)$$

Geometrically, the model $N^{(1)}(x)$ corresponds to the parabolic function $N^{(1)}(x)$ being shifted by $(1+r)/2r$ in the negative x-direction (Fig. 4.1). This similarity explains the qualitative transference of results from model (2.5) to (4.3). However, these differences between the two models also highlight the fact that the constructions of the discrete and iterated logistic models were separate, if related, endeavours (Section 2.3).

Li and Yorke (1975, p. 985) show that the sequence of points N_0, N_1, $N_2, \ldots, N_n$ generated by (4.2) is equivalent to following the fate of one point N_0 on the polynomial functions of the iterated logistic model (4.3), that is to the sequence N_0, $N^{(1)}(N_0)$, $N^{(2)}(N_0), \ldots, N^{(n)}(N_0)$. Depending on whether the point N_0 is a periodic or aperiodic point of the iterated model, the resulting discrete sequence will be periodic or aperiodic, respectively. Furthermore, since the periodic points on the Cantor set Λ become increasingly sparse, the likelihood of choosing a starting point N_0 that leads to an aperiodic sequence is approaching one. The discrete logistic model (2.4) could be constructed in the same way from the shifted iterated logistic model $N^{(n)}(x)$, which qualitatively has the same dynamics as the iterated logistic model. Accordingly, Li and Yorke (1975) show that, if the appropriate iterated model is discretized in this way, then the resulting discrete model will display the aperiodicity that is a defining criterion of stochastic chaos (Section 3.4.4).

Further evidence for the inclusion of a form of discretization into the antecedent of Conditional 1 is provided by Guckenheimer et al. (1977). They link the iterative model (2.5) to the corresponding discrete model (4.2) in two different ways. First, by considering plots of the discrete model's points N_x against points at time steps shifted by k steps, that is N_{x+k}. The continuous function $N_{x+k}(N_x)$, here viewing N_x as the range of all possible values on the interval, then corresponds to the polynomial function $N^{(k)}(x = N_x)$ of the iterated model. The fixed points of this function $N_k(x = N_x)$ are periodic points of order k in the corresponding discrete model, of course. Guckenheimer et al. (1977, pp. 105–107) are therefore able to recover Li's and Yorke's (1975) result that the structure of the periodic points of the iterated logistic model implies that there are an infinite number of values N_x on the discrete model (4.2) that will be periodic and a large number of points that will not be, namely any of those values that will never be periodic points of the iterated logistic model $N^{(n)}$ (x). All of these results qualitatively transfer to the discrete logistic model and the shifted iterated model, respectively.

Guckenheimer et al. (1977, pp. 110–111) then present another way of constructing the discrete model (4.2): as the sequence of points on the first iteration $N^{(1)}(x)$ of the iterated model that one obtains by successively projecting ("bouncing") back and forth between this function and the identity function $f(x) = x$. The same process of construction for the discrete logistic model (2.4) and the shifted iterated model (4.3) is shown in Fig. 4.1. It is apparent from Fig. 4.1 that this process of construction through discretization not only recovers the aperiodic behaviour of the discrete model in the chaotic phase but also the asymptotic and periodic behaviour for smaller r values as well as the "extinction" regime for larger r values. For the case of $r = 4$, the extinction regime will be absent and the aperiodic behaviour continues infinitely (e.g. Hirsch et al., 2004).

These investigations with the iterated logistic model therefore provide evidence that the following conditional holds true in the discrete logistic model:

Conditional 3. Logistic (quadratic) form $\wedge$ Discretization
$$\text{by } \frac{\text{Point tracing}}{\text{Projecting}} \ldots \rightarrow \text{Stochastic chaos.}$$

Conditionals which posit a form of non-linearity and a form of discretization as conjoint sufficient conditions for chaos will be denoted as type 2 conditionals (Section 4.3.1).

Both Li and Yorke (1975) and Guckenheimer et al. (1977) use the iterated logistic model to find out more about a version of the discrete logistic model. In particular, they are able to show that, if the underlying equations of the model are logistically quadratic and the model can be reconstructed by one of the processes of discretization presented in this section, then the discrete model will show stochastic chaos (Conditional 3). Importantly, this implies that certain factors are ruled out as necessary conditions for stochastic chaos: both Li and Yorke (1975) and Guckenheimer et al. (1977) constructions of apparently stochastic sequences show that no error source or other probabilistic introduction of randomness is needed to obtain stochastic chaos in the discrete logistic model. This does formally not rule out that such factors may also be sufficient conditions for stochastic behaviour. However, since the analytic framework used here is based on the acceptance of conditionals by practitioners (Suarez, 2013), it is not necessary to rule out all other possible sufficient conditions to claim that Conditional 3 is the one accepted by practitioners.

The case study of the evaluation of the iterated logistic model is therefore a good illustration of the investigative use of horizontal models in chaos theory and of their role in determining conditionals to be transferred from related vertical models, in particular. The case study also illustrates the fact that the definition of chaos applied to the model, Devaney chaos, highlights – in its requirements – exactly the features of the model that are important for these investigations.

4.3   EVALUATION OF VERTICAL CHAOTIC MODELS

In this section, I will follow up the case study of the evaluation of the discrete logistic model (Section 4.2.1) by a general discussion of how vertical chaotic models are evaluated. As in my analysis of the case study, I will be using the three-step conceptualization of the evaluation of a conditional C → B to be transferred from the model to the target system as an analytic framework. Each step in the evaluation process – the determination of the conditional to be transferred; the determination of the existence of the predicted kind of chaos in the target system; and the determination of model faithfulness – will be discussed in detail in Section 4.3.1, 4.3.2 and 4.3.3, respectively.

4.3.1   Determining the Conditional to be Evaluated

In the three-step analytic framework for the evaluation of vertical chaotic models, the evaluation process begins with an explicit specification of the conditional C → B to be transferred from the model to the target system. As we have already seen in the case study of the evaluation of the logistic lineage (Section 4.2), such conditionals can be variably construed: different authors are interested in different chaotic behaviours B and assign different sufficient conditions C to these behaviours. However, I will show that the large number of different conditional generated in this way can be sorted into two general classes, depending on the antecedent C of the conditional. In this section, I will also again emphasize the investigative role of horizontal models during the determination of conditionals C → B in related vertical models (as illustrated in Section 4.2.2).

Variability in the Chaotic Behaviour B
The behaviour B in the conditional C → B to be transferred from a vertical chaotic model (or investigated in a horizontal model) is, of course, chaos. However, due to the large number of different chaos definitions (Section 3.4), the behaviour described by this label can fulfil different combinations and embodiments of the criteria for chaos. As illustrated in Sections 2.3.1 and 4.2, this variability can mean that models from the same lineage are diagnosed with different kinds of chaos and, consequently, also assigned different conditionals C → B during their evaluations. I argued in Sections 3.4 and 3.5 that the different definitions used in this scenario are often the ones that serve the investigative role of a particular model best. Additionally, the fact that there is considerable overlap between the extensions of the different chaos definitions (Section 3.4, Table 3.3) means that a single model can be diagnosed with several different kinds of chaos and, consequently, undergo several evaluations with different conditionals C → B.

For example, in (Section 4.2.1), we have seen that Hassell et al. (1976) use the definition of stochastic chaos in their evaluation of the discrete logistic model. In a later review article on the detection of chaos in insect populations, which the authors explicitly describe (p. 2) as a follow-up to Hassell et al. (1976), Hastings et al. (1993, p.C4) state:

> The simplest and most intuitive definition of chaos is extreme sensitivity to initial conditions [on an attractor].

The authors then specify one embodiment of SDIC, namely the existence of positive Lyapunov exponents (p. 6) as their formal requirement for the diagnosis of chaos. Since they also require the existence of an attractor (Section 3.3.2), the chaos definition used here is one in terms of the existence of a strange attractor (Section 3.4.5). Similar preambles determining the specific definition of chaos to be used can be found in virtually all papers on the evaluation of chaotic models (e.g. the papers collected in Ott et al., 1994).

Two Types of Sufficient Conditions C

As illustrated in Section 4.2, the antecedents C of the conditionals C → B to be transferred from vertical chaotic models, or investigated in horizontal models, can also be variably defined. However, as apparent in Conditionals 1–3, there exists more agreement on the sufficient conditions C for the occurrence of chaos than on the definitions of the different chaotic behaviours B. In particular, I will argue that two general types of sufficient conditions can be identified and that this classification can also be used to classify the conditionals C → B as such.

There seems to be a consensus among chaos scientists that nonlinearity is a sufficient condition for chaos in all of the investigated models (e.g. Hassell et al., 1976; Ott et al., 1994; Tsonis and Elsner, 1989; Lorenz, 1993; Chernov and Markarian, 2006). In Conditionals 3 and 2, the form of non-linearity was quadratic; other forms of non-linearity include mixed linear terms as found in the Lorenz equations (Section 2.4), modulus operations (tent map), singular changes of directions (billiard dynamics), non-linear topological operations (Smale's horse-shoe). A form of non-linearity is therefore part of both types of sufficient conditions C assigned to the conditionals C → B to be transferred from chaotic vertical models.

Type 1 Conditionals

As we have seen in Section 4.2, other sufficient conditions for chaos are usually not as explicitly specified by practitioners but can be reconstructed from the investigative context. Conditional 2 thereby seems to be an indicative example of a class of conditionals that posits a form of non-linearity and a form of iteration as sufficient conditions for a particular kind of chaotic behaviour. I will call this type of conditionals type 1 conditionals:

Conditional 4 (Type 1). Form of non-linearity ∧ Form of iteration → Chaotic behaviour.

We will encounter further examples of type 1 conditionals in the case study of the evaluation of the Lorenz lineage (Section 4.5). The forms of non-linearity and iteration used in type 1 conditionals are model and representation dependent. However, some general forms can be discerned in the literature. The two iterative non-linear processes identified as sufficient conditions for Devaney chaos in the iterated logistic model – iterative construction of a fractal set and the topological operation of "folding" (Conditional 2) – have been identified in other models with similar kinds of chaos as well (e.g., for review, Smith, 1998, chapters 3 and 6). In particular, such processes can also be identified in the dynamics of the iterated Lorenz model: in the iterative narrowing of the invariant areas of the first return map and in the folding of trajectories into the attractor (Section 2.4.2).

For multi-dimensional models, the topological "kneading sequence" of "stretching", "folding" and "compressing" has been linked to the occurrence of chaos in many chaotic models, particularly those displaying kinds of chaos similar to Devaney chaos (Section 3.4.1). Smith (1998, p. 179) even suggests that, since iterative "stretching" and "folding" is at least a sufficient condition for many kinds of chaos, the identification of such a sequence could be used as a criterion for the definition of chaos. Under this interpretation of the "kneading sequence", that is as a sufficient condition for certain kinds of chaos, the seminal work by Smale (1967) can be viewed as constituting investigative work with a horizontal model, namely the horseshoe map. The result of this investigative work is the establishment of the kneading sequence as a non-linear process that – in conjunction with iteration – can be part of the antecedent of a type 1 conditional. The fact that the kneading sequence is a sufficient condition for some kinds of chaos has been used in the determination of type 1 conditionals in other chaotic models and has also been generalized to some degree (for a philosophical review of the importance of this sequence, Smith, 1998, Chapter 6; for a technical review, Hirsch et al., 2004, Chapter 16).

Type 2 Conditionals
In (Section 4.2), we have seen that there also exists a second type of chaotic conditional. These conditionals also posit non-linearity as a sufficient condition for chaos but then posit discretization as an additional condition. We recall: discretization in Section 2.3.2 has been introduced as the use of, or the switch to, a set of discrete independent variables, which distinguishes discretization and iteration since the latter acts on continuous entities. While the usefulness of this distinction might be somewhat context dependent, it appears to be a useful structuring devise in my analysis of modelling in chaos theory. Similarly, I find it useful to clearly distinguish conditionals C → B that contain discretization in the antecedent C from type 1 conditionals (Conditional

4), which contain iteration in the antecedent. I will call this second class of conditionals type 2 conditionals:

Conditional 5 (Type 2). Form of non-linearity ∧ Form of discretization → Chaotic behaviour.

Conditional 3 is an example of a type 2 conditional. In the case of the discrete logistic model, strong evidence that this conditional holds true, that is for the fact that if the quadratic equation is discretized in a particular way, then the resulting model will show stochastic chaos, is provided by the investigative work with the iterated logistic model (Section 4.2.2).

Further evidence for the fact that discretization and non-linearity can be sufficient conditions for chaos in one-dimensional maps similar to the discrete logistic model is provided by Tsonis (1992, pp. 132–133). The author points out that many such maps, including the discrete logistic maps, can be represented as modulus maps (through the choice of a particular coordinate system). A modulus map can be visualized as "moving the binary point [of a given number] to the right . . . and dropping the integer part" (p. 133). Accordingly, once a value in a modulus model is represented as a real number, that is a number without recurrent decimal sequence, the modulus operation will generate aperiodic values and hence the aperiodicity required by the definition of stochastic chaos.

Smith (1998, p. 163–164) points out that the philosophical implications of this account are currently unclear: this account of stochastic chaos seems to indicate that "facts about chaotic randomness depend on for example, facts about how most reals are random". In my account such questions about the dependency of stochastic chaos on the representation of numbers become somewhat less pressing since I can interpret works like Tsonis (1992, pp. 132–133) non-problematically as evidence towards the importance of discretization as a sufficient condition for stochastic chaos. It is notable that such investigations also involve a kind of horizontal modelling through generating a modulus model from the initial model under investigation.

Correlations Between Types of Conditional and Kinds of Chaos

My analysis here seems to indicate that Devaney chaos is usually part of a type 1 conditional (and not of a type 2 conditionals) and that stochastic chaos is usually be part of a type 2 conditional (and not of a type 1 conditional). Some of the investigative work on the sufficient conditions for different types of chaos can be viewed as supporting these correlations: for example, it is difficult to see how the modulus operation identified by

Tsonis (1992, pp. 132–133) could lead to the kind of properties necessary to fulfil the criteria of Devaney chaos.

However, for the majority of chaos definitions such correlations have not been established. Furthermore, due to the fact that most chaos definitions overlap in their extensions and are model specific in their embodiments, the same chaotic behaviour can be part of several different conditionals, including conditionals of different kinds. For example, the definition of chaos as mixing (Section 3.4.2), which has been designed to have a maximal extension and to capture those aspects shared by the majority of other chaos definitions, could be substituted both for "Devaney chaos" in Conditional 2 as well as for "stochastic chaos" in Conditional 3. Mixing therefore has sufficient conditions in the iterated logistic model that are different from those it has in the discrete logistic model. The determination of a chaotic conditional C → B is therefore highly model and definition specific and usually requires a significant amount of investigative work. Furthermore, as I will discuss in the next paragraphs, the nature of the model is not necessarily a guideline towards the type of conditional practitioners wish to establish: I will argue below that a significant part of the modelling efforts in chaos theory can be interpreted as attempts to determine type 1 conditionals as the conditionals to be transferred from discrete numerical models.

Determining Conditionals in Numerical Models
The role of numerical models in chaos theory has been a subject of ongoing debate (e.g. Kellert, 1992; Schurz, 1996; Smale, 1998). I will argue in this paragraph that two different approaches to the evaluation of such models can be discerned: (i) the approach that I will consider the mainstream viewpoint among practitioners, which aims to determine type 1 conditional as the conditionals to be transferred from these models; and (ii) an approach to the determination of the sufficient conditions for chaos in numerical models that is based on the assumption that global unpredictability should be taken to be the defining feature of chaos (Section 3.3.5). This latter approach seems have received more attention in the philosophical literature on chaos than in the technical literature. As foreshadowed in Section 3.3.5, I will argue that the latter interpretation of chaos is beset by fundamental difficulties, which will be revealed clearly in the analysis of the possible sufficient conditions for global unpredictability in numerical models.

We recall: numerical models are models whose governing set of equations is continuous but which had to be discretized to allow for numerical integration. The discrete Lorenz model (Section 2.4.1) is an example of a numerical model. Numerical models are therefore discrete models. However, it is clear from the construction of the model that discretization is a fictional aspect, which reflects the mere technical necessity of numerical integration. It is notable that this is different for the discrete logistic model (Section 2.3.2): in this model, discretization is not fictional but a true representation of the discrete generational dynamics of the target system.

The fact that discretization is a fictional aspect in numerical models has significant consequences for the evaluation of these models. According to my definition of model faithfulness (Definition 1) and the requirements for the acceptance of the transfer of a conditional $C \rightarrow B$ laid out by Suarez (2013), a behaviour B is not modelled faithfully if the antecedent C contains fictional parts. Numerical models therefore cannot model chaotic behaviour B faithfully if discretization is seen as part of the sufficient conditions for chaos in this model. As I will illustrate in the case study of the evaluation of the discrete logistic model (Section 4.5), the usual approach to the evaluation of these models by practitioners is to present evidence towards the existence of a type 1 conditional in these models (i.e. towards the fact that part of the dynamics of the model can be represented as an iterative, non-linear process on a set of continuous trajectories and that this process leads to the occurrence of chaotic behaviour) and thereby avoid having to consider discreteness a necessary condition for the chaotic behaviour under investigation. It will become apparent in the case study that such investigations usually require a large amount of investigative work with related horizontal models. This appears to be one of the reasons why the establishment of a type 1 conditional in the discrete Lorenz model has even been named by Smale (1998) as one of the twenty most important mathematical problems to be solved in the twenty-first century. Similarly, a detailed investigation of the sufficient conditions for chaos in a numerical model appears to have only been conducted for the discrete Lorenz model.

The possible influence of discretization on the behaviour of a numerical model is an important question in the evaluation of these models in other fields of science as well, of course. Accordingly, general techniques have been developed to quantify the effects discretization might have on the trajectories of a numerical model. One of the most prominent techniques is that of shadowing, which has been applied to some chaotic models and which I will therefore briefly outline here. A discrete trajectory of points is said to be shadowing a continuous trajectory, if it has been established that the two trajectories do not deviate from each other by more than a small distance (variably defined). Proving shadowing for continuous trajectories that cannot be expressed analytically, for example, for hypothetical solutions to sets of continuous differential equations that cannot be solved analytically, is technically difficult and has to be accomplished for each set of equations individually (e.g. for review, Palmer, 2000, pp. 241–256).

Accordingly, investigations of shadowing properties have only been undertaken for the most prominent chaotic numerical models and have mostly been focussed on the discrete Lorenz model (e.g. Palmer, 2000, Chapter 11). For the majority of numerical models in chaos theory, no shadowing theorems have been derived. Palmer (2000, Chapter 11) reviews existing work and derives some new results on shadowing in the discrete Lorenz model and the Henon map. For both models, the author is able to establish the existence of some shadowing periodic orbits, albeit not dense periodicity (p. xii). The shadowed orbits also have a form of SDIC: in some cases, different periodic orbits can result from only slightly different initial conditions. The fact that shadowing results exclusively pertain to the periodic properties of the discrete Lorenz model is not surprising since this technique's proper field of applications is simple periodic dynamics: it is therefore "not directly applicable" (Palmer, 2000, p. xii) to chaotic models. This might be one of the reasons why investigations of the shadowing properties of chaotic numerical models, which could be used to provide evidence towards the existence of type 1 conditionals in these models (e.g. by showing that the shadowed trajectories can be generated through a kneading sequence), appear to be rare compared to investigations relying on related horizontal models (e.g. as illustrated in Section 4.5).

While much of the investigative work on numerical models in chaos theory can be interpreted as attempts to establish the existence of a type 1 conditional in these models, there also exists some works, mostly of philosophical nature, that appear to argue for the existence of a different conditional in these models. These accounts are related to the definition of chaos as global unpredictability (Section 3.3.5). We recall: this account states that chaos defined as global unpredictability is a consequence of the amplification of small, random truncation errors during the numerical integration of models with SDIC. In

Section 3.3.5 I have argued that the definition of chaos as global unpredictability is undesirable due to the following drawbacks of this definition: (i) it seems to be to specifically geared towards numerical models; and (ii) it does not adequately reflect the fact that global unpredictability can be a consequence of several combinations of other criteria for chaos. Here, I will point out another problem with this account: namely, that it forces us to view the existence of both discretization and an error source as sufficient conditions for global unpredictability and, if defined as synonyms, chaos.

In the context of the evaluation of vertical models and the possible transfer of a conditional C → B to a target system, it is obvious that this interpretation entails a particular version of a type 2 conditional:

Conditional 6. Form of SDIC ∧ Form of discretization ∧ Truncation/ numerical errors → Global unpredictability (Chaos).

On the one hand, due to the inclusion of truncation or other numerical errors in the antecedent, Conditional 6 appears to be exclusively applicable to numerical models. On the other hand, the fact that both truncation/ numerical errors and discretization are part of the antecedent of this conditional implies that global unpredictability, and therefore chaos, cannot be modelled faithfully by numerical models. Some authors who follow this particular interpretation of chaos seem to view the inability to be modelled faithfully as an important conceptual characteristic of chaos (e.g. Chernov and Markarian, 2006, p. 48; Schurz, 1996). However, as we have already seen in the case studies of the models in the logistic lineage, there are some paradigmatic chaotic models which model chaos faithfully. Assuming that chaos cannot be modelled faithfully from the outset also leaves open the question of whether it can be a property of a natural system at all. This has led to some speculations on the question of whether only systems which need to be solved numerical, that is which cannot be integrated analytically, can be chaotic (e.g. Chernov and Markarian, 2006, p. 48; Schurz, 1996). However, this account would again imply that none of the logistic models nor any of the other horizontal models used in chaos theory would be labelled "chaotic". Similarly, viewing an absence of model faithfulness as indicative of chaos seems to be incompatible with the aims of much of the modelling work underdone in chaos theory, which, as I have described above, can be interpreted as the gathering of evidence towards the existence of conditionals that would allow the faithful numerical modelling of chaos. My analysis here therefore seems to provide further evidence towards the point of view that defining chaos as global unpredictability has philosophically problematic implications (e.g. Batterman, 1993; Werndl, 2009c).

4.3.2   Determining the Existence of Chaos

In Section 4.3.1, I argued that there are several ways in which the conditional C → B to be transferred from a vertical chaotic model can be established. In this section, I will discuss the second step of the three-step conceptualization of model evaluation (Section. 4.1), that is the determination of the existence of chaos in the target system. This section will therefore focus on the second part of the conditional, the chaotic behaviour B, and its detection in data from natural systems. I will show that the main difficulty encountered by practitioners during this step is the need to ensure that the dynamical (pre-)criterion of determinism is fulfilled, that is ensuring that the dynamics underlying a given natural target system are deterministic (Section 3.3.1).

As described in Section 4.3.1, the behaviour B in the conditional C → B is the type of chaos to be predicted by the conditional. Due to the large number of co-existing chaos definitions, the precise specification of the requirements for a diagnosis of this behaviour can be different for each evaluation. However, in Section 3.3, I have argued that virtually all chaos definitions can be decomposed into combinations of embodiments of five primary criteria: the dynamical criterion of determinism (Section 3.3.1); and the four phenomenological criteria of transitivity (Section 3.3.2), periodicity (Section 3.3.3), aperiodicity (Section 3.3.4) and SDIC (Section 3.3.5). Completing this second step of the evaluation process therefore involves showing that the target system fulfils the specific combination of embodiments of the criteria required by the chaos definition used to define the behaviour B. While the first step of the evaluation process, that is the determination of a conditional C → B that holds true in the model and should be considered for transference to the target system, is also often completed for horizontal models (e.g. Section 2.3.3), this second step of the evaluation process is specific to vertical models.

For the astronomical models discussed by Suarez (2013), detecting the behaviour B in their target systems is not problematic and is therefore viewed as a precondition rather than a separate conceptual step in the evaluation process. However, in chaos theory, formally determining whether chaos is present in a natural system has proven difficult for two reasons. First, determining whether a system fulfils the embodiments of the four phenomenological criteria of chaos often require knowledge of the behaviour of a system in specific phase space regions and for different initial conditions. Given that experimental data is often limited in both of these regards, specific techniques are used to extrapolate from incomplete (compared to the diagnostic requirements) data sets. In the next paragraph, I will discuss in detail the most prominent of these diagnostic tools, phase-space reconstruction. Second, ensuring that the dynamical criterion of determinism is fulfilled requires drawing conclusion about the system's dynamics from phenomenological data, that is from data about its behaviour rather than its underlying mechanisms. The development of techniques to do so has been one of most prolific subfield of chaos theory and will also be discussed below.

Phase-Space Reconstruction as a Diagnostic Tool

The most common technique for analysing potentially chaotic data from a natural target system is phase-space reconstruction (e.g. for a review and a reprint of seminal papers, Ott et al., 1994, Chapters 5, 6 and 7). This technique does not assume an underlying model (in contrast to the data fitting method used in the evaluation of the discrete logistic model, Section 4.2.1) but is based on the correlation of an observed time-series with parameter- or variable-shifted versions of itself. By generating several extrapolated time-series and investigating their correlation with each other, general properties of the system's behaviour in phase-space can be estimated: including phase-space correlation (Section 3.3.2); local (Section 3.3.4) and global (Section 3.3.5) prediction parameters; the spectrum of periodic orbits (Section 3.3.3); the existence and dimensions of attractors (Section 3.4.5); and the system's Lyapunov coefficients (Section 3.3.5).

These properties can then be used to check whether a model fulfils the requirements of a given definition of chaos, that is whether it possesses the required embodiments of a given combination of criteria. This can usually be done straightforwardly for the four phenomenological criteria, but requires additional assumptions in the case of the dynamical criterion of determinism. The most prominent approach to using such phenomenological data as evidence for the presence of underlying deterministic dynamics, the route-to-chaos approach, will be discussed below.

Koperski (1998, p. 639) views phase-space reconstruction as bottom-up model construction (Section 2.2.1). In this author's view, models which show aperiodicity and/or SDIC (and are therefore chaotic according to definitions based on these criteria) cannot be evaluated by a bit-by-bit comparison of the models' output with the behaviours of their target systems. Koperski (1998) maintains that this difficulty in comparison has lead experimental chaos scientists to largely eschew top-down modelling (p. 645):

> Instead of starting with a mathematical model and making predictions about the trajectory of the system, the experimentalist starts with a time-series in order to reconstruct the phase-space.

I agree with Koperski (1998) that phase-space reconstruction does not technically depend on the prior existence of a top-down model. However, in my opinion, interpreting this technique as a model construction activity is incorrect for three reasons (which can be illustrated, for example, in the papers collected in Ott et al., 1994 and Robertson and Combs, 1995).

First, as described in Section 2.2.1, more recent work on the use of vertical models in science suggests that model evaluation generally does not consist of the simple bit-by-bit comparison described by Koperski (1998, p. 625). Accordingly, what he takes to be the canonical version of top-down modelling appears to be a simplified version of the process, which incorrectly (in my opinion) asserts that the top-down construction of models makes no reference at all to the empirical context.

Second, in the vast majority of experimental papers in chaos theory, the analysis of natural data is only undertaken because a previously constructed top-down model suggests that the system might display chaos, that is in my framework, that a conditional C → B to be transferred to the target system can be determined in the model. Sometimes the underlying model is so well known (e.g. the discrete Lorenz model or discrete logistic model) that no renewed derivation of the model is deemed necessary; this might give the misleading impression that the phase space reconstruction has been performed in isolation.

Third, the end-product of a phase-space reconstruction appears to be not a model but a diagnosis: based on the information gained from the reconstruction, the natural system under consideration is judged to be chaotic (according to a given definition) or not.

Accordingly, I maintain that viewing phase-space reconstruction as a diagnostic tool, which allows the researcher to gather qualitative evidence towards the existence of chaos in a given target system, is a better interpretation of the actual use of this technique. One of the advantages of the framework of model evaluation used in this book, that is, the three-step conceptualization of the (possible) transference of a conditional C → B (Sections 2.2.1 and 4.1), is a shift of focus away from the requirement of exact quantitative agreement of behaviours (which would put too high a demand on models' convergence to their target systems in most fields of science). Instead, at this stage in the evaluation process, the scientist merely has to decide whether there is enough evidence to convince him that chaotic behaviour of the desired kind is present.

My view of phase space-reconstruction as a diagnostic tool is also compatible with Smith's (1998, Chapter 5) idea that a notion of approximate resemblance of chaotic behaviour in models and in their target systems can be established by emphasizing geometrical similarities. In my framework of model evaluation, such similarities can be seen as "good enough" indicators that chaos (of a given definition) is present in the system. Here, too, phase-space reconstruction is best viewed as an efficient diagnostic tool to reveal such similarities.

Diagnosing Determinism: The Route-to-Chaos Approach

As discussed in Section 3.3.1, for many kinds of chaos, the four phenomenological criteria are not sufficient to distinguish chaotic behaviour from the behaviour of indeterministic systems. This is particularly apparent for stochastic chaos (Section 3.4.4), whose phenomenological state, by definition, is indistinguishable from an appropriate comparison state of a truly stochastic system (i.e. of a Bernoulli process). Accordingly, data from the chaotic regime of the target system of a chaotic model is not sufficient to show that the underlying dynamics of the system are deterministic. In order to make sure that the system fulfils this criterion for chaos, data from other phases of the system's development need to be taken into account. This approach to diagnosing determinism in a natural system is often called the route-to-chaos approach (e.g. Smith, 1998, p. 102). Since all definitions of chaos require determinism as a (pre-)criterion, the following discussion of this technique applies to all kinds of chaos diagnosed in vertical models.

Underlying the route-to-chaos approach is the idea that it is the development of a system before it reaches its chaotic phase that distinguishes chaotic systems from stochastic systems. The term development can thereby apply to both a development along the dependent variables as well as a development along changes of a parameter in the model. There appear to be two general classes of route-to-chaos approaches: one geared towards models and systems that are subject to Sarkovskii's theorem (Theorem 1); and one applicable to all models and systems.

The first class of route-to-chaos approaches utilizes the period-doubling regime that precedes the chaotic regime in models like the two logistic models (Section 2.3). Since the existence of this regime is another direct consequence of Sarkoviskii's theorem, one can assume that, in models to which Theorem 1 applies, period doubling will always be present at some stage. In contrast, there is no reason to expect that a truly stochastic system would undergo the same run-up to the display of stochastic behaviour. Identifying such a period doubling parameter regime in natural systems could therefore be used to distinguish stochastic chaos in these systems from noise. Period doubling has been found in several laboratory studies on the existence of chaos in natural systems (e.g. for a collection of studies, Cvitanovic, 1986, Chapter 2).

The route-to-chaos approach based on the identification of period-doubling is well supported theoretically by results on the generation of chaos in models like those in the logistic lineage (Section 4.2). However, period-doubling can only be expected for models in which Sarkovskii's theorem is supposed to hold or which are closely related to such models. As described in Section 4.3.1, for many models, it is not clear whether their dynamics can be represented in a way that allows the application of the theorem, that is as an iterative process which eventually generates periodic points of prime period 3, or whether they can be related to such a model.

Accordingly, there exist a second class of route-to-chaos approaches to diagnosing determinism (e.g. Smith, 1998, Chapter 4; Tsonis, 1992, pp. 238–244). These approaches are based on the assumption that phase-space correlation (Section 3.3.2), local (Section 3.3.4) and global (Section 3.3.5) prediction parameters develop differently in truly stochastic systems and in chaotic systems. Since most chaotic systems do not instantly reach their chaotic phase (even if period doubling is not expected) or possess some localized non-chaotic parameter regimes, these parameters will decrease with ongoing development of the system, e.g. if phase-space reconstruction is used, the correlation between time-series shifted only a few time steps away from each other will be better than the correlation between time-series shifted by many time steps. In contrast, truly stochastic time-series have consistently low correlation and prediction parameters. This contrast has also be phrased in terms of long-term and short-term predictability: the former should be low in both chaotic and stochastic systems, while the latter should be significantly higher in chaotic systems.

The question of whether such route-to-chaos approaches reliably distinguish underlying deterministic dynamics from stochastic ones is not fully resolved yet (e.g. Pool, 1989; Ott et al., 1994, Chapters 10–11). In particular, due to restrictions on the quality and scope of observational data, the "route-to-chaos" behaviour can usually not be detected in natural systems with the same clarity as it can be seen in models. Furthermore, for the second class of approaches, which relies on the identification of decay in correlation and prediction parameters, the possibility that there are some kinds of probabilistic noise that show a similar variability cannot be categorically ruled out. Both of these restrictions apply more strongly in the case of natural systems, which is also reflected in the current state of research: diagnoses of determinism, and therefore of chaos, in laboratory systems have been made with relative certainty while the presence of chaos in natural systems is still debated (e.g. Smith, 1998, Chapter 8). Solving these problems has and continues to be a focus of research activity in chaos theory (e.g. Cvitanovic, 1986; Ott et al., 1994). This concentration of research activity on the second step of the model evaluation process appears to be a distinguishing epistemic features of chaos theory as a scientific field.

4.3.3   Determining Model Faithfulness

In this section, I will discuss the third step in my three-step conceptualization of model evaluation in chaos theory: the determination of model faithfulness. We recall from Section 2.2.1: according to Suarez (2013), once the existence of the behaviour B has been established in both the model and the target system, scientists will be willing to accept the transference of a conditional C → B from the model to the target system if the antecedent C can be shown to not contain any fictional aspects of the model. Alternatively, an aspect B is modelled faithfully if it can be viewed as part of such a non-fictional conditional C → B (Definition 1). An evaluation of model faithfulness therefore entails a comparison of the model to the target system, with the aim of deciding which aspects of the posited sufficient conditions C exist in both the model and the target system, that is which conditions C are fictional and which are not.

Neither scientists (e.g. Cvitanovic, 1986; Ott et al., 1994; Hirsch et al., 2004) nor philosophers (e.g. Kellert, 1993; Smith, 1998; Koperski, 1998) have so far explicitly discussed the faithfulness of chaotic models. As illustrated in the case study of the discrete logistic model (Section 4.2.1), studies with vertical chaotic models usually include some discussion of particular factors in the model or the target system that might have influenced the results obtained, but there is usually no explicit identification of fictional and non-fictional conditions. Given that the three-step framework is a means of rational reconstruction rather than a direct description of scientific practice, the fact that not all conceptual steps are given equal attention by practitioners, or are equally well articulated in the literature, is not concerning. Furthermore, my analysis of the evaluation process within this framework allows me to identify two primary reasons for the fact that model faithfulness (or similar concepts) have so far not been given much attention in chaos theory.

First, determining the existence of chaos in the relevant target systems – the second step in the evaluation process – has proven to be particularly difficult for chaotic models. As we have seen in 4.3.2, these difficulties can be traced back to the need to infer the fulfilment of the dynamical criterion of determinism from knowledge about the phenomenology of the system. These difficulties derive from the general structure of the definitions of chaos and they are therefore specific to the field of chaos theory. It is apparent from both the scientific (e.g. Cvitanovic, 1986; Ott et al., 1994; Hirsch et al., 2004) as well as the philosophical literature (e.g. Kellert, 1993; Smith, 1998; Koperski, 1998) that the majority of intellectual energy and attention has been focussed on resolving these difficulties. This focus on the diagnosis of chaos in the relevant target systems supports Suarez' (2013) point of view that the detection of the behaviour B in both the model and the target system is a minimal requirement for the transference of a conditional.

Second, the determination of conditionals C → B in vertical chaotic models is in itself a laborious process and usually requires significant investigative work with related horizontal model (Section 4.2.2, 4.3.2 and 4.5). For the vast majority of models, this step of the evaluation process is still incomplete, which means that either intellectual effort is focussed on furthering the relevant understanding of the underlying model (e.g. as illustrated in the case studies in Section 4.2) or that a discussion of the sufficient conditions for chaos is omitted all together (e.g. as illustrated by the majority of models presented in Lorenz, 1993; Robertson and Combs, 1995).

Likewise, the aspects of chaos theory that have been most interesting to philosophers are related to the definition of chaos (Chapter 3); the consequences of chaos (e.g. pertaining to predictability, as discussed in

Sections 3.3.4 and 3.3.5); and the difficulties in determining the existence of chaos in nature (4.3.2). In the philosophical literature, model faithfulness has therefore also received little attention. This lack of attention could also be due to the fact that, in the existing philosophical frameworks, it would be difficult to discuss model faithfulness in a generalized way, that is any such discussion would have to be tied to specific models and their technical evaluation. However, the establishment of the two different types of chaotic conditionals C → B (Section 4.3.1) now allows me to provide a more generalized discussion of model faithfulness. In the following two paragraphs, I will therefore discuss the model faithfulness of type 1 and type 2 conditionals, respectively.

Model Faithfulness and Type 1 Conditionals
In type 1 conditionals, a form of iteration and a form of non-linearity in the governing equations are viewed as sufficient conditions for chaos. As discussed in Section 4.3.1, this type of conditional is often linked to Devaney chaos (Section 3.4.1) and other kinds of chaos, whose extensions overlap with that of Devaney chaos (Section 3.4). Since Devaney chaos most often is a feature of horizontal models, that is models without a target system, type 1 conditionals are also often associated with this class of models. Accordingly, for many models in which a type 1 conditional has been determined to hold true, an evaluation of model faithfulness is not necessary. This was illustrated in the case study of the iterated logistic model (Section 4.2.2 and Conditional 2).

However, as described in Section 4.3.1, type 1 conditionals have also been determined (indirectly, through work with related horizontal models) for numerical models. We recall: despite their discrete nature, these models cannot model chaotic behaviour that is part of a type 2 conditional faithfully. Therefore, the establishment of an alternative set of sufficient conditions in the form of a type 1 conditional has been the aim of a significant amount of modelling activity in chaos theory. The work-intensive nature of such investigation means that convincing evidence for the existence of a type 1 conditional currently only exists for very well-known models, like the discrete Lorenz model (Section 2.4.1). In Section 4.5, I will discuss the evaluation of this model in detail.

Assuming that a type 1 conditional has been determined in a numerical model, an evaluation of model faithfulness would then require a detailed comparison of the features of target system's dynamics to those forms of non-linearity and of iteration required in the conditional. Once these two aspects have been identified as non-fictional, the model could be judged to model a given chaotic behaviour faithfully. To my knowledge, no such evaluation has been conducted so far. In the case of the discrete Lorenz model, where a type 1 conditional has been determined, the detection of chaos in the relevant target systems, that is weather systems, has proven difficult (e.g. Tsonis and Elsner, 1989). Beyond the characteristic difficulties of the detection of chaos in a natural system (Section 4.3.2), the determination of model faithfulness for the Lorenz model is further compounded by the fact that the model's dynamics have been greatly simplified in comparison to the highly complex, multi-scale dynamics of its target systems, so that an evaluation of the model against its target systems is generally more complex than for models without such a discrepancy in descriptive richness.

Model Faithfulness and Type 2 Conditionals
The use of type 2 conditionals in chaos theory was demonstrated in the case study of the discrete logistic model (Section 4.2.1), in which Conditional 3 holds true. In type 2 conditionals, the antecedent consists of the conjunction of a form of non-linearity and a form of discretization (Section 4.3.1). In the case of the discrete logistic model, one can straightforwardly show that the model represents stochastic chaos faithfully since neither the quadratic form of the model's equations nor the discretization of the independent variable x appear to be fictional. May (1974, p. 645) makes it clear that the quadratic form of the model's equations is actually a conservative representation of non-linearity, that is the restriction to second-order polynomials of N_x is justified by an order of magnitude estimation, which shows that higher order terms exert little influence on the function. Accordingly, this part of the antecedent in Conditional 3 can be judged as (at least in essence) non-fictional. Furthermore, May (1974, p. 645) stresses that the model's target systems are populations with non-overlapping, that is discrete, generational dynamics. Accordingly, the discreteness of the independent variable x is also a true representation of the target system's dynamics and hence not a fictional aspect. Conditional 3 therefore does not contain any fictional conditions and should be considered for transference to the target system, that is to insect populations with discrete dynamics.

The evaluation of the discrete logistic models shows that chaotic behaviour that is type 2 conditionals can be modelled faithfully. However, this is only the case if the discrete nature of the model is not fictional. (Non-linearity appears to be a robust true feature of virtually all chaotic models.)

As discussed in Section 4.3.1, this is not the case for numerical models: the discrete nature of these models is a technical artefact of the numerical integration. Numerical models are therefore examples of models in which chaos cannot be faithfully modelled as part of a type 2 conditional – at least not if the conditional refers to the discretization associated with numerical integration. Accordingly, in what I take to be the mainstream account of the evaluation of numerical models in chaos theory (Section 4.3.1), practitioners try to avoid having to consider discreteness as a condition for chaos in these models by determining other sufficient conditions; this is usually accomplished by the determination of a type 1 conditional to hold true in these models (Section 4.3.1 and 4.5).

4.4   Evaluation of Horizontal Chaotic Models

In this section, I will explicitly discuss the evaluation of horizontal chaotic models. As described in Section 2.2.2, such an evaluation does not involve the possible transference of a conditional C → B to a target system. Instead, the evaluation of horizontal chaotic models involves (i) an investigation of their own mathematical properties and (ii) their use in the investigation of other vertical models in the lineage of models they belong to. The latter function (ii) can be interpreted in my three-step analytic framework for the evaluation of vertical models as a contribution to the determination of the conditional C → B to be transferred from a related vertical model (Section 4.3.1). The evaluation of horizontal models in chaos theory can therefore usually not be separated from that of other models in their lineage. Accordingly, the two aspects of such evaluations have already been touched upon in my earlier discussion of the evaluation of vertical chaotic models (Section 4.3). Here, I will examine them explicitly.

4.4.1   Investigation of Mathematical Properties

In Section 4.3.1, I have argued that investigations of the mathematical properties of horizontal chaotic models are not fundamentally different from investigations of the mathematical properties of vertical chaotic models: both can be conceptualized as the determination of a conditional C → B, which specifies the chaotic behaviour B observed in the model and a set of sufficient conditions C for this behaviour. This is illustrated in the case study of the evaluation of the iterated logistic model (Section 4.2.2), where a significant part the investigation of its mathematical properties can be summarized (roughly) as the determination of Conditional 2.

The first step of the three-step conceptualization of model evaluation can therefore be completed for both vertical and horizontal models. This overlap in the evaluation procedures is not surprising: at this stage in the evaluation process no knowledge of or reference to a model's target system is needed.

4.4.2   Investigative Use

The remaining two steps of the evaluation process for vertical models, the determination of the existence of chaos in the target system (Section 4.3.2) and the determination of model faithfulness (Section 4.3.3) are specific to the evaluation of vertical models; horizontal models, which have no target systems, cannot undergo these steps. Instead, I maintain that horizontal chaotic models are used as investigative tools in the determination of conditionals in related vertical models, that is they are used as tools in the first step of the evaluation process for other models in their lineage. As described in Section 4.3.1, determining which conditional C → B holds true in a chaotic model is a difficult and highly model-specific process. The construction and manipulation of horizontal models to aid the investigation of a given vertical model appears to be an integral part of this evaluative step.

The case study of the evaluation of the logistic models (Section 4.2) illustrates this role of horizontal models: here, the horizontally constructed iterated model is used to clarify the relationship between the two chaotic logistic models and to trace back the stochastically chaotic behaviour of the discrete model to both the choice of a discrete representation as well as the underlying quadratic dynamics (Section 4.2.2; Conditional 3). The investigative work with the iterated logistic model can therefore be conceptualized as the determination of a type 2 conditional in the discrete logistic model, which can then be considered for transference to the latter model's target system (Section 4.2.1). A similar case study will be presented in Section 4.5, where I will discuss the evaluation of the discrete logistic model with the aid of the iterated logistic model. In this case study, the conditional determined in the discrete Lorenz model is a type 1 conditional. As described in Section 4.3.1, the investigation of conditionals in multi-dimensional numerical models, like the discrete Lorenz model, is even more complex than the investigation of such conditionals in one-dimensional maps.

Horizontally constructed models appear to be ubiquitous in chaos theory. Not all of these are directly spun off from vertical models: some models, like the Baker map or the tent map, appear to have been constructed purely to explore their mathematical properties. Similarly, results from investigations with horizontal models are often transferable to models that are not part of the model's lineage. For example, results from investigations with Smale's horseshoe map (e.g. Smale, 1967), which has been constructed as a rough topological representation of the dynamics of some planetary systems, have provided guidance towards the determination of sufficient conditions for chaos in a large range of other models (e.g. Devaney, 1989, pp. 139–147).

My analysis of the evaluation of horizontal chaotic models confirms the two main claims made by Bokulich (2003), who first discovered this class of models: it (i) provides further evidence for the investigative use of these models and it (ii) supports her assertion that – given the lack of scholarly attention focused on them so far – horizontal models are surprisingly important in some scientific fields. Through the case studies and my reconstruction of the general evaluation of models in chaos theory, I have been able to determine the investigative functions of horizontal chaotic models more specifically as the determination of chaotic conditionals $C \rightarrow B$. While this reconstruction might not be the only conceptualization of the role of horizontal models in chaos theory, the investigative use and the importance of these models in the field appear to be a solid result of my analysis.

The case study of the logistic models also illustrates an auxiliary epistemological function of horizontal models in chaos theory: these models play a particular role in making chaos theory an interdisciplinary field by constituting epistemic bridges between the natural sciences and mathematics. This will also be apparent in the detailed case study of the evaluation of the lineage of Lorenz models (Section 4.5), which connected meteorologists and mathematicians in the same way that the lineage of logistic models brought together mathematicians and biologists. The interdisciplinary nature of chaos theory is also noted in the socio-historical analysis of the field's development by Aubin and Dalmedico (2002). I hope to have pinpointed its epistemic origin more precisely to the interplay of horizontal and vertical models.

4.5   SMALE'S 14TH PROBLEM

In this section, I will return to the case study of the lineage of Lorenz models (Section 2.4) and discuss attempts to solve Smale's 14th problem. While there exist several other evaluative studies of these models (e.g. for review, Hirsch et al., 2004, Chapter 14), the posing and solving of Smale's 14th problem are arguably the most significant and well-known contributions to the evaluation of this lineage of models. The problem itself is part of a list of mathematical problems, which are to be addressed in the twenty-first century, compiled by Smale (1998, p. 15):

> Is the dynamics of the ordinary differential equations of Lorenz that of the geometric Lorenz attractor of Williams, Guckenheimer, and Yorke?

Smale (1998) then clarifies that the question aims to establish whether the system of Lorenz equations (Section 2.4.1), as an example of a system describing a physical phenomenon, will lead to similar chaotic behaviour as the horizontally constructed iterated Lorenz model, or the horseshoe map he himself investigates in Smale (1967, p. 15), displays:

> An answer to this problem would be a step in establishing foundations for the field of applied chaotic dynamical systems.... Geometric, structurally stable, chaotic attractors in dynamics are in my paper [Smale, 1967]. But these did not arise from any physical system.

In the terminology used in this book, the "geometric Lorenz attractor" is identical to the iterated Lorenz model (Section 2.4.2).

The following detailed (albeit completely non-technical) discussion of Smale's 14th problem and the accepted solution to this problem will illustrate two issues discussed previously: (i) the construction and use of a horizontal model as a tool to determine a conditional C → B to be transferred from a related vertical model to its target system (Section 4.4); and (ii) the particular difficulties which arise during the evaluation of numerical models (Section 4.3.3). With respect to aspect (ii), it will also become apparent that the evaluation of the Lorenz lineage is a good illustration of investigations aiming at the determination of a type 1 conditional in a numerical model.

4.5.1   Chaotic Conditionals in the Lorenz Models

It is obvious from the phrasing of Smale's 14th problem and from the later solution provided by Tucker (2002) that the question is not just whether the discrete Lorenz model, which provides a numerical solution to the Lorenz equations, can be diagnosed with the same kind of chaos as the iterated Lorenz model but also whether the discrete model represents chaos faithfully, that is in my analytic framework, whether the chaotic behaviour in the discrete model can be seen as part of a conditional without fictional parts in the antecedent.

Thereby, the chaotic conditional that holds true in the iterated Lorenz model is well known since this model is constructed with the sole aim of determining a set of conditions under which a particular chaotic behaviour occurs (Section 2.4.2). We recall: the iterated Lorenz model consists of a set of explicit phase-space constraints on the behaviour of the continuous trajectory of an unspecified function, which lead to an iterative, non-linear folding of this trajectory onto a butterfly-shaped attractor. The dynamics induced by these constraints therefore lead to a behaviour of the model that both is visually similar to that of the discrete Lorenz model as well as shares many chaotic properties with the iterated logistic model (and the horseshoe map). Using the results from Section 4.3.1, we can now see that Guckenheimer et al. (1977) establishes the existence of a type 1 conditional (Conditional 4) in the iterated Lorenz model:

Conditional 7.   Non-linear folding of trajectories ∧ Iteration
$\rightarrow$ Strange attractor
Devaney chaos   . . ..

As indicated in Conditional 7, the iterated Lorenz model has been found to be chaotic in several different ways (Table 3.3) so that the consequent in Conditional 7 can vary. However, the sufficient conditions for these behaviours appear to be well established through the deliberate construction of the model.

In contrast, the sufficient conditions for chaos in the discrete Lorenz model are less clear, for example, it is not obvious whether a type 1 or type 2 conditional holds true in the model (Hirsch et al., 2004, p. 303, Smith, 1998, p. 183). This difficulty is due to the fact that the Lorenz model is a discrete model of a continuous set of equations, that is a numerical model. As discussed in Section 4.3.1, as the discrete nature of these models is an artefact of the numerical integration algorithm, chaos as part of a type 2 conditional cannot be modelled faithfully by numerical models. Accordingly, practitioners (in general) try to rule out discretization as a necessary condition for chaos in the model; instead, they aim for the determination of a type 1 conditional. The construction of the iterated Lorenz model can already be interpreted as part of an investigation with these aims: this construction shows that non-linear folding of trajectories and iteration can be sufficient conditions for a chaotic behaviour that is similar to the one observed in the discrete Lorenz model.

Accordingly, in my analytic framework, Smale's 14th problem can be reinterpreted as posing the question of whether Conditional 7 or, at least a conditional of the same type, can be shown to hold true in the discrete Lorenz model or in another model whose trajectories also constitute solutions to the Lorenz equations. As I will discuss in the next paragraphs, the accepted solution to the problem by Tucker (2002) consists of the construction of a semi-numerical model of the Lorenz equations in which the effect of discretization has deliberately been eliminated.

4.5.2   Construction of the Rigorous Lorenz Model

The most successful attempt at solving Smale's 14th problem is Tucker (2002). Tucker's methodology is usually viewed as the design of an improved version of the discrete Lorenz model by using a more rigorous numerical integration mechanism. I maintain that his work is better interpreted as the addition of a third, horizontally constructed, model to the lineage of Lorenz models (Section 2.4). The model constructed by Tucker (2002), like the discrete Lorenz model, provides a solution to the Lorenz equations. However, in contrast to the discrete Lorenz model it is not constructed vertically from these equations as governing theory. Instead, a closer look at Tucker (2002) reveals that the newly constructed model, which I will call the rigorous Lorenz model, is a "hybrid" model, which contains both parts adopted from the discrete Lorenz model as well as parts adopted from the iterated Lorenz model.

A crucial feature of the rigorous Lorenz model is its construction as a semi-numerical model: the model's domain is divided into two parts, a boxed shaped region around the origin (the cusp of the butterfly), in which its trajectories are determined analytically, and the remainder of the domain, in which its trajectories are integrated numerically. In order to be able to perform an analytic integration within the area close to the origin, it is assumed that the dynamics in this part of the domain are linear, which is an assumption also made in the iterated Lorenz model (Section 2.4.2). Tucker (2002) transfers this assumption to the rigorous Lorenz model and linearizes the flow in a box-shaped region centred around the origin. In order to minimize the impact of this alteration, he also changes the coordinate system inside this box, employing a coordinate system "which is virtually linear" (p. 60). Accordingly, the dynamics of the rigorous model near the origin are very similar to those of the iterated model and can also be viewed as iterative, that is whenever a trajectory enters the analytic box an iteration is the performed that determined the linear progression of this trajectory through the box.

Outside the box around the origin, the trajectories are determined by a numerical integration algorithm similar to the one used in the discrete Lorenz model. However, Tucker (2002) improves this numerical integration mechanism by using "rigorous numerics" (p. 60). In particular, Tucker's (2002) integration algorithm uses interval arithmetic rather than single point computations. Therefore, rather than computing single trajectories, the algorithm returns a bounding region, which contains all possible trajectories resulting from the forward integration of all points between the upper and lower limit of a directly rounded value. The algorithm runs on an adaptive mesh to keep the interval size relatively small throughout the time of integration. The aim of this rigorous integration mechanism is to eliminate the influence of discretization, and truncation errors in particular, on the model's behaviour.

This construction through a combination of parts from the two existing models strongly suggests that Tucker's (2002) rigorous Lorenz model should be viewed as a third, horizontally constructed, model in the Lorenz lineage. In particular, the use of both iterated and discrete dynamics in the rigorous Lorenz model implies that it cannot be seen as merely an updated version of the discrete Lorenz model. Similar to the other horizontal model in the Lorenz lineage, the iterated Lorenz model, the rigorous Lorenz model has an explicit investigative function: it is constructed with the deliberate aim to model chaos as part of a conditional $C \rightarrow B$ which does not contain fictional aspects in the antecedent and which can be transferred from the rigorous model to the target system of the discrete Lorenz model, for example, the Lorenz equations. While the rigorous model is therefore not directly informative about the modelling of chaos in the discrete Lorenz model, it is informative about the target system of this model.

4.5.3   Behaviour of the Rigorous Lorenz Model

Tucker (2002) then performs the necessary numerical and analytic integrations of the rigorous Lorenz model and analyses its behaviour. He shows that the model possesses an attractor (Section 3.4.5), which is described in the following way (p. 58):

T]he Lorenz attractor is just as the geometric [iterate] model predicts: it contains the origin and thus has a very complicated Cantor book structure...

This implies that the trajectories of the model are topologically transitive on the attractor (Section 3.3.2) and that "a tiny blob of initial values rapidly smears out over the entire attractor, as observed in numerical experiments" (p. 58), that is the trajectories exhibit a form of SDIC (Section 3.3.5). Since Tucker (2002) defines an attractor as chaotic precisely when these two criteria, transitivity and SDIC, are fulfilled (p. 114), the rigorous Lorenz model is judged to be chaotic. Accordingly, Tucker (2002) shows that the rigorous Lorenz model can be diagnosed with at least one kind of chaos that has also been found in the iterated and the discrete Lorenz model (Table 3.3).

4.5.4   Evaluation of the Rigorous Lorenz Model

It is apparent from Tucker's (2002, p. 60) justification of the rigorous numerical integration mechanism and the use of the analytic integration near the origin that these features are meant to eliminate the influence of discretization on the behaviour of the rigorous model. Accordingly, Tucker (2002) not only shows that the rigorous, the discrete and the iterated Lorenz model can be diagnosed with the same kind of chaos but that this type of chaos cannot be part of a type 2 conditional in the iterated rigorous model. Given the similarity of the dynamics near the origin between the iterated and rigorous Lorenz model, it seems like that a version of Conditional 7 should hold true in the latter model as well. Then, the fact that chaos in the rigorous model is part of a type 1 conditional implies that chaos is modelled faithfully in this model and that this chaotic conditional should be considered for transference to the model's target system, which, as described above, is actually the target system of the discrete Lorenz model. As described in (Section 4.3.3), an actual evaluation of the Lorenz model against a natural system has so far not taken place.

Computerized proofs, like the one provided by Tucker (2002), have been subjected to general criticism, questioning their ability to stand on equal footing with analytic proofs. In particular, it has been asserted that implementations that involve a large amount of code and long simulation times are practically not fully verifiable (e.g. Viana, 2000; McEvoy, 2013). Similarly, the complexity of the model and the lack of shadowing theorems for the discrete and rigorous Lorenz model mean that it is difficult to verify the exactness of the rigorous model's solution to the Lorenz equations. Nevertheless, Tucker (2002) has generally been accepted as a solution to Smale's 14th problem by the community of chaos scientists; most importantly by Smale himself (e.g. Hirsch et al., 2004, p. 304).

Tucker's (2002) methodology appears to be very much in line with the general investigative use of horizontal models in chaos theory. In this case, the model is used to be directly informative about the target system of another model in its lineage, without being constructed from any considerations about this target system itself. However, as described above, the construction of the rigorous model is horizontal. Furthermore, the importance of Tucker's (2002) result lies not the occurrence of chaotic behaviour, which was already pre-empted by the behaviour of the discrete Lorenz model, but in the (rigorous) conditions under which it occurs. Accordingly, this semi-numerical model can be seen as contributing to the determination of a chaotic conditional in a similar way as seen during the evaluation of the lineage of logistic models (Section 4.2).

4.6   CONCLUSION

In this chapter, I discussed the evaluation of chaotic models. The analytic framework used for this discussion was a three-step framework based on Suarez' (2013) account of the transference of conditionals C $\rightarrow$ B from vertical models to their target systems. In this framework, the three main steps in the evaluation of vertical chaotic models can be conceptualized as: (i) the determination of a conditional C $\rightarrow$ B to be transferred from the model to the target system; (ii) the determination of the existence of the chaotic behaviour B in the target system; and (iii) the determination of model faithfulness according to Definition 1. Since the framework is intended as a tool for rational reconstruction, I do not claim that it is descriptively accurate in any amount of historical detail or that it is normatively prescribing. However, in this chapter, I hope to have demonstrated that the framework is useful in clarifying the process of model evaluation in chaos theory and in highlighting the most important conceptual issues of this process.

In particular, in Section 4.3.1, I showed that the coexistence of a number of chaos definition (Section 3.4) leads to a corresponding coexistence of many conditionals C $\rightarrow$ B to be evaluated, whereby the behaviour B can correspond to different descriptions of chaos. However, these conditionals can be classified into two categories according to the types of sufficient conditions for chaos they posit (Section 4.3.1): type 1 conditionals have a form of non-linearity and form of iteration as their antecedent (Conditional 4); while type 2 conditionals have a form of non-linearity and a form of discretization as their antecedent (Conditional 5). It also became apparent that the investigative role of horizontal models in chaos theory (Section 2.2.2) can often be conceptualized as aiding in the determination of such a conditional in a related vertical model (Section 4.4).

During this analysis of the determination of conditionals in chaos theory, I also showed that the definition of chaos as global unpredictability (Section 3.3.5) necessitates the positing of sufficient conditions for chaos that are neither of type 1 nor of type 2. Instead, the antecedent of the corresponding conditional includes both discretization and the existence of a fundamental source of uncertainty (Conditional 6). I then argued that the necessity to accept such conditionals highlights the problematic nature of this interpretation of chaos. In particular, Conditional 6 implies that chaos can never be modelled faithfully and leaves its status as a natural property uncertain. My analysis therefore offers further support for the point of view that global unpredictability is not a suitable definition of chaos (e.g. Werndl, 2009c; Batterman, 1993).

The determination of the existence of the behaviour B in the target system is viewed as a precondition in Suarez' (2013) framework. However, for the evaluation of chaotic models, this process warrants representation as a separate step since it is technically difficult and has traditionally received the most attention of all three steps (Section 4.3.2). In particular, technical difficulties arise from the need to ensure the fulfilment of the dynamical criterion of determinism from phenomenological data about the system. The methodology that has been developed to solve this difficulty are "route to chaos" -techniques, which differentiate between truly probabilistic and chaotic systems by using information from the non-chaotic phases of the latter.

The last step in the evaluation of chaotic models is the determination of model faithfulness, that is the determination of whether the antecedents of the conditionals to be transferred contain no fictional parts. In (Section 4.3.3), I showed that for both type 1 as well as type 2 conditionals, this process is highly model specific. In particular, a behaviour B that is part of a type 2 conditional will only be modelled faithfully if the target system itself is discrete, for example, as is the case for the discrete logistic model, whose target systems are population with discrete generational dynamics. A particular class of models for which this is not the case, that is which are discrete models of continuous target systems are numerical models. I then argued that the evaluations of numerical models in chaos theory therefore usually involve the determination of a type 1 conditional and that a significant part of the modelling activity in chaos theory can be interpreted as attempts to rule out discretization as a necessary condition for chaos in numerical models. The case study of the evaluation of models in the Lorenz lineage provides a good illustration of such investigations. It also highlights their work intensive nature.

In Section 4.4, I emphasized the importance and investigative role of horizontal models in chaos theory. My results on the construction and use of these models in chaos theory support Bokulich's (2003) results on the use of these models in semi-classical physics. I have also been able to specify the main investigative function of horizontal models in chaos theory more specifically: namely, as an investigative tool in determining the conditional C → B to be transferred from a related vertical model.

My analysis in this chapter is supported by two case studies: the evaluation of the lineage of logistic models (Section 4.2) and the evaluation of the lineage of Lorenz models (Section 4.5). Both case studies illustrate the construction and use of horizontal models during the evaluation of a related vertical model very well. Furthermore, the case study of the logistic models is an example of the relatively unproblematic determination and evaluation of type 1 and type 2 conditionals. Nevertheless, this evaluation required a considerable amount of investigative work: Hirsch et al. (2004, p. 336) state that the "work of hundreds of mathematicians" has been necessary to understand the logistic models. The second case study of the evaluation of the Lorenz lineage also illustrates this difficult and work intensive nature of the evaluation of chaotic models. Here, the evaluation of these models has even been included in a list of mathematical problems to be solved in the twenty-first century (Smale, 1998). Thereby, the evaluation of the discrete Lorenz model is an example of the evaluation of a numerical model in chaos theory. As described above, the case study illustrates the tendency of practitioners to search for type 1 conditionals in these models. It also illustrates the technical complexity of these investigations and the need to rely on the horizontal construction of additional models and on computerized proofs.

CHAPTER 5

Conclusion

Abstract I will provide a summary of the books content. In the three main chapters of this book, I analysed the construction, diagnosis and evaluation of models in chaos theory. Using the method of rational reconstruction, the book aimed to analyse each of the three stages of modelling in detail. To my knowledge, this is the first such modelling-centred analysis of chaos theory.

Keywords chaos · conclusion

During this analysis, I developed two novel analytic frameworks: a novel approach to the analysis of chaos definitions based on a view of these definitions as twofold decomposable into combinations of five core criteria and into different formal embodiments of these criteria (Chapter 3); and an inferential framework for the evaluation of chaotic models based on the transference of conditionals between model systems and target systems (Chapter 4, Suarez, 2013).

In the following paragraphs, I will summarize my discussions of each of the three stages of modelling in chaos theory and highlight the main conceptual results obtained for each of these stages.

L.C. Zuchowski, *A Philosophical Analysis of Chaos Theory*, New Directions in the Philosophy of Science, DOI 10.1007/978-3-319-54663-6_5

5.1   Construction of Models in Chaos Theory

In Chapter 2, I reviewed relevant material on the construction of scientific models, and developed the main structure of my framework for the analysis of the evaluation of chaotic models. I maintained that there are two kinds of model construction in chaos theory: vertical construction and horizontal construction.

Vertical models are constructed from top-level theory and bottom-level empirical data (Section 2.2.1). Their evaluation, therefore, consists of a comparison between the model and its target system. However, rather than focusing on a bit-by-bit comparison between these two entities, I based my own analytic framework on the inferential account of model evaluation developed by Suarez (2013). This account conceptualizes the evaluation of vertical models as the evaluation of a conditional C → B to be transferred between the model and the target system. Underlying this framework is the assumption that scientists are not only interested in the occurrence of a certain behaviour B but also in the sufficient conditions C for this behaviour. As I hope to have demonstrated in the subsequent chapters, this assumption is borne out in the case of chaos theory. In the context of this framework, I introduced a notion of model faithfulness (based on the initial conceptualization of this notion by Bolinska, 2013), which states that a behaviour B is modelled faithfully if the antecedent C in an appropriate conditional C → B does not contain any fictional parts, that is if the behaviour B is conditionally isolated from any fictional effects arising from the model's construction.

I then (Section 2.2.2) maintained that there exists a second important class of models in chaos theory: horizontal models. Horizontal modelling was first discussed by Bokulich (2003) but has not received much further attention to date. Horizontal models are not constructed from governing theories and empirical data but are mathematical variations of existing models. The horizontal construction of models can therefore lead to lineages of mathematically related models. Horizontal models have no target systems and their evaluation does therefore not consist of a comparison to a natural system. Instead, following Bokulich (2003), I maintain that horizontal models have an investigative function. I consider this function to be highly subject specific. In the case of chaos theory, as I demonstrated in Chapter 4 of this book, this function consists of the investigation of the mathematical properties of other, vertical models in a given lineage.

In Sections 2.3 and 2.4, I illustrated the construction of such lineages of chaotic vertical and horizontal models in two case studies: that of the logistic lineage and of the Lorenz lineage, respectively. The logistic lineage includes the vertically constructed discrete logistic model (Section 2.3.2) and the horizontally constructed iterated logistic model (Section 2.3.3). The Lorenz lineage includes the vertically constructed discrete Lorenz model (Section 2.4.1), the horizontally constructed iterated Lorenz model (Section 2.4.2) and a third horizontally constructed model, the rigorous Lorenz model, which was introduced later in the book (Section 4.5). All four models discussed in this chapter have achieved iconic status in chaos theory. They were also used as case studies to illustrate my analyses throughout the book.

5.2   DIAGNOSIS OF MODELS AS BEING CHAOTIC

In Chapter 2, I discussed the diagnosis of models as being chaotic. This entailed the introduction of a novel framework for the rational reconstruction of chaos definitions (Section 3.1).

This analytic framework is based on the assumption that chaos definitions can be decomposed into requirements of different combinations of five core criteria for chaos (Section 3.3): the dynamical criterion of determinism (Section 3.3.1); and the four phenomenological criteria of transitivity (Section 3.3.2), periodicity (Section 3.3.3), aperiodicity (Section 3.3.4) and SDIC (Section 3.3.5). Each of these criteria can be conceptualized as a similarity category, which contains a number of more concrete, formal embodiments of the concept.

I argued that this mode of analysis highlights the fact that different chaos definitions can be (i) geared towards the specific technical requirements of different models through the choice of appropriate embodiments and (ii) highlight characteristics that are important in the investigative use of a model through the choice of particular combinations of criteria. The co-existence of many different chaos definitions therefore appears to be not a symptom of deep conceptual disagreements among practitioners but rather a consequence of the complex interactions of different chaotic models. Furthermore, the framework allows an interpretation of determinism as a criterion for the diagnosis of chaos, which, I maintain, is a good representation of the use of this concept in chaos theory (Sections 3.3.1 and 4.3.2).

I further demonstrated the benefits of my analytic framework by analysing five prevalent definitions of chaos (Section 3.4): Devaney chaos

(Section 3.4.1); (the definition of chaos as) mixing (Section 3.4.2); (the definition of chaos in terms of) positive Lyapunov exponents (Section 3.4.3); stochastic chaos (Section 3.4.4) and (the definition of chaos in terms of) strange attractors (Section 3.4.5). Each of these definitions was shown to require different combinations and embodiments of the five criteria. It became apparent that the extensions of the different definitions overlap considerably, which further mitigates any conceptual conflicts between them. I also briefly discussed the definition of chaos as global unpredictability but argued that global unpredictability should be seen as consequence rather than a definition of chaos.

My results were illustrated in the case study of the diagnoses of the models in the logistic lineage (Section 2.3.1). The discrete logistic model was initially found to be stochastically chaotic (Section 3.2.2) while the iterated logistic model was primarily diagnosed with Devaney chaos (Section 3.2.3). The main conceptual results, as outlined above, could also be demonstrated in the case study.

5.3   Evaluation of Chaotic Models

In Chapter 4, I discussed the evaluation of chaotic models. I further developed the framework for the analysis of model evaluation (Section 2.2.1) by adapting it specifically to the evaluation of chaotic models. I identified three conceptual steps in the evaluation process (Section 4.1): (i) the determination of a conditional C → B to be transferred from a model to a target system; (ii) the determination of the existence of chaos in the target system and (iii) the determination of model faithfulness (Definition 1).

In Section 4.3.1, I discussed the determination of chaotic conditionals C → B in detail. The coexistence of many different chaos definitions means that many different conditionals can be constructed through the use of variable definitions of the behaviour B. However, I then maintained that, in these conditionals, there are usually only two types of sufficient conditions C for chaos: conditionals of type 1 posit a form of non-linearity and a form of iteration as sufficient conditions; while conditionals of type 2 posit a form of non-linearity and a form of discreteness as sufficient conditions. The existence of these two types of conditionals was demonstrated in the case study of the logistic models.

In Section 4.3.2, I discussed the determination of the existence of chaos in the relevant target systems. It became apparent that this step in the evaluation process is particular difficult to complete for chaotic models. In particular, difficulties arise from the need to determine the fulfilment of the criterion of determinism from phenomenological data about systems that, in their chaotic phases, are (by definition) phenomenologically indistinguishable from stochastic ones. This difficulty appears to be a unique epistemic characteristic of chaos theory.

In Section 4.3.3, I discussed the determination of model faithfulness for chaotic models. The case study of the logistic lineage was used to show that chaos can be modelled faithfully as part of both type 1 and type 2 conditionals. However, I then discussed two cases in which the determination of model faithfulness is more problematic. Numerical models, that is, discrete models of continuous sets of equations, can only model type 1 conditionals faithfully. I illustrated the necessity to find such conditionals in numerical models in the case study of the evaluation of the Lorenz lineage (Section 4.5). I also showed that a definition of chaos as global unpredictability implies that chaos can never be modelled faithfully since the appropriate conditional for this definition includes the existence of an in-principle error source as a sufficient condition for chaos.

In Section 4.4, I discussed the evaluation of horizontal models in chaos theory. Horizontal models are not evaluated against a target system. Instead, their primary epistemic role is an investigative one: they are used as investigative tools in the evaluation of related vertical models in their lineage. Using the three-step analytic framework for the evaluation of these models, I could further specify this investigative function as aiding in the determination of the conditionals to be transferred from these related vertical models. The investigative role of horizontal models was demonstrated in two case studies: that of the evaluation of the logistic lineage (Section 4.2) and that of the evaluation of the Lorenz lineage (Section 4.5).

5.4 INTERPLAY OF MODELS AS A CHARACTERISTIC FEATURE OF CHAOS THEORY

My analysis of modelling in chaos theory seems to indicate that the complex interplay of different models is one of the most distinguishing features of chaos theory as a scientific field. In particular, I hope to provide a clear exposition of the importance and epistemic role of horizontal models in chaos theory. As demonstrated in Chapter 2, the interaction of different models also influences the definition of chaos: the coexistence of many different chaos definitions appears to be directly related to the existence of many different models with different epistemic functions. My work here therefore seems to indicate that what might appear to be signs of conceptual inhomogeneity are actually consequences of the complex interactions of different modelling activities.

BIBLIOGRAPHY

H. D. I. Abarbanel, R. Brown, J. J. Sidorowich, and L. Sh. Tsimring. The analysis of observed data in physical systems. *Reviews of Modern Physics*, 65: 1331–1392, 1993.

K. T. Alligood, T. D. Sauer, and J. A. Yorke. *Chaos: An Introduction to Dynamical Systems*. Springer, New York, 1997.

D. Aubin and A. Dahan Dalmedico. Writing the history of dynamical systems and chaos: Longue duree and revolution, disciplines and cultures. *Historia Mathematica*, 29: 273–339, 2002.

D. M. Bailer-Jones. Scientists' thoughts on scientific models. *Perspectives on Science*, 10: 275–301, 2002.

J. Banks, J. Brooks, G. Cairns, G. Davis, and P. Stacey. On Devaney's definition of chaos. *The American Mathematical Monthly*, 99: 332–334, 1992.

R. W. Batterman. Defining chaos. *Philosophy of Science*, 60: 43–66, 1993.

A. Berger. *Chaos and Chance: An Introduction to Stochastic Aspects of Dynamics*. Walter de Gruyer, New York, 2001.

A. Bokulich. Horizontal models: From bakers to cats. *Philosophy of Science*, 70:609–627, 2003.

A. Bolinska. Epistemic representation, informativeness and the aim of faithful representation. *Synthese*, 190: 219–234, 2013.

M. Boshernitzan, G. Galpertin, T. Kruger, and S. Troubetzkoy. Periodic orbits are dense in rational polygons. *Transactions of the American Mathematical Society*, 350: 3523–3535, 1998.

N. Cartwright. *How the Laws of Physics Lie*. Clarendon, Oxford, 1983.

N. Chernov and R. Markarian. *Chaotic Billiards*. American Mathematical Society, New York, 2006.

G. Contessa. Scientific models and fictional objects. *Synthese*, 172: 215–229, 2010.

© The Author(s) 2017
L.C. Zuchowski, *A Philosophical Analysis of Chaos Theory*, New Directions in the Philosophy of Science, DOI 10.1007/978-3-319-54663-6

J. M. Cushing. *Integrodifferential Equations and Delay Models in Population Dynamics*. Springer, Berlin, 1977.

P. Cvitanovic. *Universality in Chaos*. Adam Hilger, Bristol, 1986.

R. L. Devaney. *An Introduction to Chaotic Dynamical Systems*. Addison Wesley, Redwood City, 1989.

J. Earman. *A Primer on Determinism*. Kluwer, Dordrecht, 1986.

R. Frigg. Models and fiction. *Synthese*, 172: 251–268, 2010.

Roman Frigg and Stephan Hartmann. Models in science. In Edward N. Zalta, editor, *The Stanford Encyclopedia of Philosophy*, Fall edition, 2012.

R. N. Giere. *Explaining Science: A Cognitive Approach*. The University of Chicago Press, Chicago, US, 1988.

P. Godfrey-Smith. Models and fictions in science. *Philosophical Studies*, 143: 101–116, 2009.

R. Gonczi and C. Froeschle. The Lyapunov characteristic exponents as indicators of stochasticity in the restricted three-body problem. *Celestial Mechanics*, 25: 271–280, 1981.

J. Guckenheimer. A strange, strange attractor. In J. E. Marsden and M. Mc-Cracken, editors, *The Hopf Bifurcation and Its applications*, pages 368–381. Springer, Berlin, 1976.

J. Guckenheimer, G. Oster, and A. Ipaktchi. The dynamics of density dependent population models. *Journal of Mathematical Biology*, 4: 101–147, 1977.

A. Hájek. Interpretations of probability. In E. N. Zalta, editor, *The Stanford Encyclopedia of Philosophy*. 2012.

M. P. Hassell, J. H. Lawton, and R. M. May. Patterns of dynamical behavior in single-species populations. *Journal of Animal Ecology*, 45: 471–486, 1976.

A. Hastings, C. L. Hom, S. Ellner, P. Turchin, and H. C. J. Godfrey. Chaos in ecology: Is mother nature a strange attractor? *Annual Review of Ecology and Systematics*, 24: 1–33, 1993.

R. C. Hilborn. *Chaos and Nonlinear Dynamics*. Oxford University Press, Oxford, 2002.

M. W. Hirsch, S. Smale, and R. L. Devaney. *Differential Equations, Dynamical Systems and An Introduction to Chaos*. Elsevier, Amsterdam, 2004.

K. Judd and L. Smith. Indistinguishable states II: The imperfect model scenario. *Physica D*, 151: 125–141, 2001.

K. Judd and L. Smith. Indistinguishable states: Perfect model scenario. *Physica D*, 196: 224–242, 2004.

S. H. Kellert. A Philosophical evaluation of the chaos theory „revolution". *PSA: Proceedings of the Biennial Meeting of the Philosophy of Science Association*, 33–49, 1992.

S. H. Kellert. *In the Wake of Chaos*. The University of Chicago Press, Chicago, 1993.

S. H. Kellert. *Borrowed Knowledge: Chaos Theory and the Challenge of Learning across Disciplines.* The University of Chicago Press, Chicago, 2008.

S. H. Kellert, M. Stone, and A. Fine. Models, chaos and goodness of fit. *Philosophical Topics*, 18: 85–106, 1990.

J. Koperski. Models, confirmation and chaos. *Philosophy of Science*, 65: 624–649, 1998.

M. Kot and W. M. Schaffer. Discrete-time growth-dispersal models. *Mathematical Biosciences*, 80: 109–136, 1986.

H. A Lauwerier. One-dimensional iterative maps. In Arun V. Holden, editor, *Chaos*, pages 39–57. Manchester University Press, Manchester, 1991.

A. Levy. Modeling without models. *Philosophical Studies*, 172: 781–798, 2015.

T.-Y. Li and J. A. Yorke. Period three implies chaos. *The American Mathematical Monthly*, 82: 985–992, 1975.

E. Lorenz. Deterministic nonperiodic flow. *Journal of the Atmospheric Sciences*, 20: 130–141, 1963.

E. Lorenz. *The Essence of Chaos.* UCL Press, London, 1993.

S. Luzzatto, I. Melbourne, and F. Paccaut. The Lorenz attractor is mixing. *Communications in Mathematical Physics*, 260: 393–401, 2005.

R. M. May. Biological populations with non-overlapping generations: Stable points, stable Cycles, and chaos. *Science*, 15: 645–647, 1974.

R. M. May. Simple mathematical models with very complicated dynamics. *Nature*, 261: 459–467, 1976.

R. M. May. Spatial chaos and its role in ecology and evolution. In S. A. Levin, editor, *Frontiers in Mathematical Biology*, pages 326–344. Springer, Berlin, 1994.

R. M. May and G. F. Osler. Bifurcations and dynamic complexity in simple ecological models. *The American Naturalist*, 110: 573–599, 1976.

C. Mayo-Wilson. Structural chaos. *Philosophy of Science*, 82: 1236–1247, 2015.

Mark McEvoy. Experimental mathematics, computers and the a priori. *Synthese*, 190: 397–412, 2013.

E. Ott, T. Sauer, and J. A. Yorke. *Coping with Chaos: Analysis of Chaotic Data and The Exploitation of Chaotic Systems.* Wiley, New York, 1994.

K. Palmer. *Shadowing in Dynamical Systems: Theory and Application.* Springer, Boston, 2000.

R. Pool. Is it chaos, or is it just noise? *Science*, 243: 25–28, 1989.

R. Robertson and A. Combs. *Chaos Theory in Psychology and the Life Sciences.* Lawrence Erlbaum, Hove, 1995.

D. Ruelle. *Chance and Chaos.* Princeton University Press, Princeton, 1991.

G. Schurz. Kinds of unpredictability in deterministic systems. In P. Weingartner and G. Schurz, editors, *Law and Prediction in the Light of Chaos Research*, pages 123–141. Springer, Heidelberg, 1996.

S. Smale. Differentially dynamical systems. i. diffeomorphisms. *Bulletin of the American Mathematical Society*, 73: 747–816, 1967.

S. Smale. Mathematical problems for the new century. *The Mathematical Intelligencer*, 20: 7–15, 1998.

P. Smith. *Explaining Chaos*. Cambridge University Press, Cambridge, 1998.

M. A. Stone. Chaos, prediction and Laplacean determinism. *American Philosophical Quarterly*, 26: 123–131, 1989.

M. Suarez. Fictionals, conditionals and stellar astrophysics. *International Studies in the Philosophie of Science*, 27: 235–252, 2013.

A. Toon. *Models as Make Believe: Imagination, Fiction and Scientific Representation*. New Directions in Philosophy of Science. Palgrave Macmillan, Basingstoke, UK, 2012.

A. A. Tsonis. *Chaos: From Theory to Applications*. Plenum Press, New York, 1992.

A. A. Tsonis and J. B. Elsner. Chaos, Strange attractors and weather. *Bulletin of the American Meteorological Society*, 70: 14–23, 1989.

W. Tucker. A rigorous ODE solver and Smale's 14th problem. *Foundations of Computational Mathematics*, 2: 53–117, 2002.

M. Viana. What's new on Lorenz strange attractors. *The Mathematical Intelligencer*, 22: 6–18, 2000.

C. Werndl. Are deterministic and indeterministic descriptions observationally equivalent. *Studies of History and Philosophy of Modern Physics*, 40: 232–242, 2009a.

C. Werndl. What are the new implications of chaos for unpredictability? *The British Journal for the Philosophy of Science*, 60: 195–220, 2009b.

C. Werndl. Justifying definitions in mathematics: Going beyond Lakatos. *Philosophia Mathematica*, 17: 313–340, 2009c.

L. C. Zuchowski. Gestalt switches in Poincaré's prize paper: An inspiration for, but not an instance of, chaos. *Studies in History and Philosophy of Modern Physics*, 47: 1–14, 2014.

INDEX

© The Author(s) 2017   137
L.C. Zuchowski, *A Philosophical Analysis of Chaos Theory*, New Directions
in the Philosophy of Science, DOI 10.1007/978-3-319-54663-6

Printed in the United States
By Bookmasters